PENGUIN BOOKS

THE OTHER SIDE OF HAPPINESS

I   k Bastian is an Associate Profess..   psychological
S   ces at the University of Melbourne. His studies have been reviewed
in leading news outlets such as *Time Magazine*, *Huffington Post*,
Economist, *New Scientist*, *Scientific American*, CNN, ABC, *Globe and
Mail*, *The New Yorker* and *New York Magazine*. Brock's pioneering
approach to research has been acknowledged with the Wegner
Theoretical Innovation Prize, and his presentations have been included
on the ABC Radio National Big Ideas podcast.

BROCK BASTIAN

The Other Side of Happiness

*Embracing a More Fearless
Approach to Living*

PENGUIN BOOKS

PENGUIN BOOKS

UK | USA | Canada | Ireland | Australia
India | New Zealand | South Africa

Penguin Books is part of the Penguin Random House group of companies
whose addresses can be found at global.penguinrandomhouse.com.

First published by Allen Lane 2018
Published in Penguin Books 2019
001

Typeset by Jouve (UK), Milton Keynes
Printed and bound in Great Britain by Clays Ltd, Elcograf S.p.A.

A CIP catalogue record for this book is available from the British Library

ISBN: 978-0-141-98210-6

www.greenpenguin.co.uk

MIX
Paper from
responsible sources
FSC® C018179

Penguin Random House is committed to a
sustainable future for our business, our readers
and our planet. This book is made from Forest
Stewardship Council® certified paper.

Contents

Acknowledgements   vi
Introduction   vii

PART ONE
Avoiding Pain   1

1. Have We Reached Peak Comfort?   3
2. The Cotton Wool Generation   23

PART TWO
Embracing Pain   47

3. Painful Pleasures   49
4. Getting Tough   75
5. Connecting With Others   96
6. Finding Focus   126
7. The Meaningful Life   148
Conclusion   171

References   176

Acknowledgements

This project would not have come together without the immense help of all the talented and patient people I am surrounded by. I would like to thank Cecilia Stein and Laura Stickney from Penguin Random House UK for their patience, for their thoughtful and thorough feedback, and for making *The Other Side of Happiness* far stronger than I could have done alone. I am also very grateful to Alexis Kirschbaum for believing in the ideas from the start. Thanks also to Kate Smith, Richard Duguid and Shoaib Rokadiya from Penguin Random House.

A big thank you to Sophie Lambert from Conville & Walsh for the huge support she has provided and being a simply fantastic agent. Also to Alex Christofi for posing the challenge of writing this book in the first place.

Many of the ideas in this book are based on an academic paper that won the SPSP Dan Wegner Theoretical Innovation Prize, which was co-written with Jolanda Jetten, Matthew Hornsey and Siri Leknes. Without the ideas, guidance, expertise and sheer brilliance of these individuals none of this might have come into being. Peter Kuppens has been an amazing collaborator in working to understand why happiness cultures are sometimes counterproductive.

Finally, a huge thank you to my ever patient and loving wife Moseni Mulemba; my two amazing daughters Makisa and Tepa; my parents Lyn and Eric; my mother-in-law Rodia for teaching me that 'life goes on'; my sister Amber; and my aunt and uncle Robyn and Graham Burke for their encouraging feedback.

Introduction

Positivity has become the new crack of the upwardly mobile, and the new pill for the downtrodden and depressed. Coaches, consultants and psychologists have been pushing the message that to live well we need to seek out the positive and reframe the negative. In today's world, feeling happy is no longer simply a state of mind; it has become a marker of mental health and success. On the flip side, pain and sadness are viewed as signals of failure and of sickness: if we are not happy then there is something wrong with us and we need to fix it. It is no wonder that the painkilling and antidepressant markets, already worth billions of dollars, continue their rapid expansion. We have come to treat even commonplace experiences of pain and sadness as pathological, as things that need to be medicated and eradicated.

Confusion arises because, despite our comfortable, first-world lives, we cannot free ourselves of pain or sadness, though we have never before had access to the kind of analgesic technology that we have today. Popping a pill to remedy physical or emotional discomfort has become the norm, yet this approach is simply not working. As pain medication gets stronger we see pain-related problems rising. As antidepressant use increases we find ourselves in the midst of a depression epidemic.

Our problem is not pain or sadness. It is how we have come to orient ourselves to these sensory and emotional events. We are sold the message that our negative or uncomfortable experiences in life are inconvenient, unnecessary, that they are of no value to us and we should avoid them at all costs.

This sentiment is well captured by the slogan of a major brand of painkillers in Australia – Panadol: 'When pain is gone, life takes its place.' It's an excellent marketing strategy: medication can return us to normal and save us from needlessly unpleasant experiences. While this statement may be apt for those who experience chronic or un-remitting pain (although these are not usually the kind of people who are popping Panadols), it casts *all* pain as counterproductive to life. With slogans such as this it is no wonder we have come to believe our lives are *supposed* to be pain free. Sold on this idea, we see happiness as a natural state, the expected equilibrium. Yet, the harder we try to reach this plateau of happiness and eradicate all suffering, the more inadequate we may be led to feel.

By devaluing and avoiding our negative experiences we leave our-selves with only one pathway to finding happiness – the pursuit of pleasure. We seek out feel-good experiences, always on the lookout for the next holiday, purchase or culinary experience. This approach to happiness is relatively recent; it depends on our capacity both to pad our lives with material pleasures and to feel that we can control our suffering. Painkillers, as we know them today, are a relatively recent invention and access to material comfort is now within reach of a much larger proportion of the world's population. These technologi-cal and economic advances have had significant cultural implications, leading us to see our negative experiences as a problem and maximiz-ing our positive experiences as the answer. Yet, through this we have forgotten that being happy in life is not just about pleasure. Comfort, contentment and satisfaction have never been the elixir of happiness. Rather, happiness is often found in those moments we are most vul-nerable, alone or in pain. Happiness is there, on the edges of these experiences, and when we get a glimpse of *that* kind of happiness it is powerful, transcendent and compelling.

Drawing on this perspective, we can begin to examine what is in many cultures today an under-explored path to well-being. Pleasant experi-ences are *necessary* but not *sufficient* to produce happiness. Experiencing pleasure is an important pathway to finding happiness, but pleasure alone will never achieve this goal. In order to find true happiness we need to embrace a more fearless approach to living, we need to approach our negative experiences on the front foot . . . we need to experience pain. By 'pain', I mean all things that are not pleasure. I mean the anxiety of a significant challenge and the loneliness of failure. I mean the sadness of a relationship breakup, or the fear of our own mortality. I also mean the physical repercussions that we may experience in a range of contexts, some of which are chosen and some of which are not. These experiences not only all 'hurt', they also trigger similar neural regions of the brain and, as I will argue, play into our overall happiness in similar ways. Drawing on recent findings, including those from my own research laboratory, and current theories within psychology, anthropology and neuroscience, I will aim to show that pain and suffering are neither antithetical to happiness nor simply incidental to it. They are *necessary* for happiness. Without pain, there is no way to achieve real happiness.

OPENING OUR EYES

One of the key reasons we fail to see the relationship between pain and happiness is that we no longer see pain clearly; we have developed 'pain blind spots'. If I asked you to stop and think about your last painful experience I bet you would think of some kind of illness, injury or significant trauma. I also bet you would think of a physical experience rather than an emotional event. This illustrates two shortcomings in how we view pain.

The first is that we have developed a narrow view of pain – we tend to limit it to physical experiences alone. This view was also held by early scientists and medical practitioners, but more recently we have learned that pain exists in the brain – this is where it is experienced. As such, we now understand pain is a psychological state that may or may not be linked to the existence of a physical injury. More importantly, this has also opened our eyes to the possibility that what we commonly refer to as 'physical pain' may share important similarities with other experiences such as negative mental states ('emotional pain')[1] or negative social encounters ('social pain')[2].

The second shortcoming is that we regard pain as a problem. We see it as intimately tied to significant physical or psychological harm, rather than as a distinct psychological state. Just because an experience is negative or hurts us does not mean it is harmful for us. This way of thinking is becoming more common. Nick Haslam, a social psychologist from the University of Melbourne, has noted that our concept of harm has been expanding to include an ever-growing list of discomforts.[3] From bullying to abuse, prejudice, trauma and addiction, these concepts have expanded over time to include even mildly unpleasant events or non-normative behaviour. Given this increasing tendency to equate discomfort with harm, it is perhaps surprising to consider that almost every experimental study investigating how we respond to negative events employs methods that are devoid of harm. This is for two reasons. First, it is illegal to cause actual harm to participants and ethics committees would never approve such studies. Secondly, it turns out that we can study the impact of physical, social or emotional pain as psychological experiences that are devoid of physical or psychological harm.

Our tendency to regard pain as a problem also limits our capacity to see it as occurring in the context of experiences we might broadly refer to as positive. This is perhaps why some women seem to forget the trauma of childbirth, or why runners or chilli lovers rarely refer to running or eating Mexican food as painful. The fact is, pleasure and pain frequently co-exist. Not only do they co-exist, but they can often be dependent on one another. Yet, we fail to see this relationship and therefore we become blind to the fact that pain contributes to positive outcomes and sometimes we may even *enjoy* these painful aspects.

Our tendency to limit pain to physical injury or to focus on its more extreme and harmful manifestations leaves us with a gap in our terminology. We have very few (if any) of what are called 'hedonically neutral' experiences in life. Even if only in small ways, our various experiences tend to be to some extent positive or negative. This is why Aristotle carved the world into broad categories of pleasure and pain. Although pleasure and pain frequently occur together, we can still categorize the various elements of every experience in these two ways. This means that whether it is a broken toe or a broken heart, the irritation of an itch or the anxiety we feel before speaking in public, the negative elements of these experiences should all be broadly considered under the category of pain. Yet, we appear to feel some resistance to this flippant use of the term. It is perhaps for this reason that more frequently we refer to these experiences as 'unpleasant'. Of course, I could have written a book about unpleasantness instead of pain. But I find this deeply dissatisfying, and I also think it is exactly this approach which has caused us to become allergic to pain. Defining a whole set of experiences based on the absence of pleasure suggests such experiences are inherently lacking; specifically, they are lacking in pleasure. Moreover, how can we talk about the benefits of pain when it is simply defined by the absence of pleasure? We need to positively identify our negative experiences, rather than define them as the absence of something better. We need a term that captures the full array of our negative experiences and draws them together under a meaningful umbrella. For this, the most accurate term is pain.

WE DIFFERENTIATE OUR PAINS MORE THAN OUR PLEASURES

If you take a moment to list all the different descriptions of pain, including all negative feelings and sensations you can think of, and then create a similar list for pleasure, you may find your list for pain will be longer than your list for pleasure. Our descriptions of emotional experiences tend to be highly differentiated at the negative end of the spectrum with many more words to describe them. This increased frequency of negative (compared to positive) descriptors is evidenced in a study which asked participants to rate an exhaustive list of emotion words, over 550 in total. This exercise revealed that 62 per cent of these words were rated as negative while only 38 per cent were rated as positive; there were more negative than positive words to describe emotional states. To provide additional evidence the researchers also asked participants to rate an exhaustive list of 555 personality traits. They replicated the same pattern, finding that people rated traits relating to emotion as negative 74 per cent of the time. That is, there were three times as many words to describe negative emotional traits as there were words to describe positive emotional traits.[4]

This same pattern is observable in the way we use the terms pleasure and pain to describe physical sensations. Paul Rozin, Professor of Psychology at the University of Pennsylvania, developed a list of pleasure and pain descriptors as follows:

- **Pain (31 words)**: deep, intense, drilling, boring, dull, sharp, aching, burning, cutting, pinching, piercing, tearing, twitching, shooting, raking, gnawing, itching, stabbing, nipping, sticking, thrusting, hard, throbbing, penetrating, lingering, fitful, radiating, bitter, pricking, biting, incising.
- **Pleasure (14 words)**: intense, thrilling, sharp, delicious, exquisite, deep, fluttering, lingering, radiating, sumptuous, breathtaking, electrifying, delicate, sweet.[5]

He found twice as many words to describe painful sensations as he did to describe pleasurable ones. Rozin even admits that in order to find a range of different pleasure descriptors he had to resort to reviewing erotic literature.

When we talk about pleasure in a broad sense, we may refer to a vast array of experiences ranging from eating chocolate to sunbathing to sex. All of these experiences can rightly be referred to as pleasurable and we have fewer terms to differentiate between these different types of pleasure. When it comes to pain, however, our self-diagnosis demands specificity. While the taste of fine wine and the enjoyment of social encounters are both considered broadly pleasurable, an aching tooth and a longing heart are wholly different.

There are good reasons why we are more sensitive to negative experiences, and therefore need to categorize them more carefully – our pleasures will rarely kill us but our pains just might. For the purpose of this book, however, let us think about pain in the same way that we think about pleasure: as a broad category that draws parallels between a vast array of negative experiences. Thankfully, there are important differences between our various negative experiences, and how they operate in our lives, but just because there are differences does not mean we cannot look for overarching patterns, links and insights.

DEFINING PAIN AND PLEASURE

As Aristotle dictated, draw a line through the middle of your world, placing all things that are painful on one side and all things that are pleasurable on the other. The first thing to note is that it is often hard to know where to put some experiences. Where should you put child-birth? This is a time of both great joy and great pain. Where should you put your trips to the gym? This is an activity we enjoy (or perhaps enjoy most once it is over), but working out at the gym inevitably causes pain.

The other thing this highlights is that we have to choose not only between significant and relatively extreme experiences, but also insignificant and relatively banal experiences. Where should you put an itch or a sense of discomfort? What about feelings of relief? A broad view of pain and pleasure should encompass both the ordinary as well as the extraordinary in order to contextualize all experiences. This allows us to understand not just our significant experiences, but *all* of our experiences from this perspective: the extent to which they involve pleasure and pain.

Overleaf I have given a broad list of types of pleasure and pain considered in this book. Indeed, some will be considered more than others due to available research (for instance, a lot of my own research has focused on physical pain). I have also loosely grouped these different, more specific pains and pleasures into subcategories of physical, emotional, social and existential. This is not intended to be exhaustive or overly technical. I have loosely aligned each pain with a corresponding pleasure, although again these things are not always diametrically opposed. The purpose of this list is thematic. It highlights that we can think about pain as encompassing a variety of negative experiences and that general similarities may exist. Whether physical, social, emotional or existential, these experiences share many commonalities and they may all lead us to respond to our worlds in comparable ways. When I am reviewing work on one type of painful experience, this list should provide a basis from which to think how that might be related to painful experiences more broadly, or to the overall contrast between pain and pleasure.

WHAT ABOUT CHRONIC PAIN?

It is important to consider the distinction between chronic, un-remitting pain and acute or transient pain. While being fearless and

Pain	Pleasure
Physical pain	*Physical pleasure*
• heat	• warmth
• cold	• coolness
• discomfort/irritation	• comfort
• hunger	• eating
• thirst	• drinking
• effort	• relaxation
Emotional pain	*Emotional pleasure*
• sadness	• happiness
• depression	• elevation
• anxiety/stress	• calmness
• fear	• excitement
• dread	• relief
• threat/challenge	• safety/overcome
Social pain	*Social pleasure*
• loneliness	• social connection
• rejection	• acceptance
• loss	• reward
• failure	• success
Existential pain	*Existential pleasure*
• fear of death	• affirmation of life
• meaninglessness	• meaningfulness
• risk	• security
• adversity	• achievement
• cost	• benefit

embracing painful experiences can directly contribute to our overall happiness in life, this link between pain and happiness is much more tenuous in cases of chronic physical pain, major depression or persistent trauma. These experiences tend to overwhelm us and are very rarely sources of joy. This does not mean, however, that this book has nothing to offer those who suffer from these negative experiences. In fact, it has a great deal to offer in terms of strategies for better management.

The potential contribution of this perspective for helping people to better cope with pain became clearer to me when I was invited to give a plenary lecture at the British Pain Society's Annual Scientific Meeting in 2016. This society was first established in 1979 by a group of anaesthetists working with people suffering from intractable pain. Over time the society grew to include a large number of health professionals ranging from doctors to nurses, physiotherapists, research scientists, psychologists, occupational therapists and other healthcare professionals. These are all people whose work is focused on finding solutions to pain, who are seeking the best ways to treat and manage it.

They specifically wanted me to talk about my own research on the benefits of pain. Although such a perspective could easily have offended people who spend their entire professional lives trying to eradicate pain, offence certainly did not characterize their reaction. Rather, these clinicians were some of the most interested and engaged people that I had spoken to, and they were very open to new ideas. Specifically, they were interested in the idea that telling patients about potential benefits might provide them with a different perspective from which to understand their pain. Framing pain in these ways, they felt, could facilitate a curious approach to pain, rather than a threat-motivated desire to avoid it. Indeed, getting people to psychologically assess their pain, rather than trying to avoid it, is one of the key strategies used by psychologists in pain management clinics.

This same approach is fundamental to the treatment of depression, anxiety and most other forms of emotional, social or existential pain. Clinicians send the message that in order to better manage these negative experiences we need to accept them, rather than struggle with them. It is this same approach which underpins our understanding of how post-traumatic growth occurs. On occasion, a person's character can grow and develop as a result of adversity and trauma – they can literally become better people. For this to happen, however, they need to be able to respond to these earth-shattering events by accepting them and accommodating them in their understanding of the world – how things are. Yet, when we view such experiences as problems that need to be solved through medical intervention, it makes it hard to accept and accommodate them. The medicalization of pain and sadness has shaped our experience of suffering; it is no longer a part of our human experience, to be expected and understood, but rather it is unnatural and counterproductive, to be treated and resolved. This disease-based model of human turmoil fails to understand pain as an integral part of what makes us human.

It was a very medical view of pain that struck me as I walked around the stalls at the meeting of the British Pain Society. There were a large number of stalls marketing the latest advances in pain management technology, from devices to medications. This diversity of interventions made it very clear to me that there was no 'silver bullet' for pain treatment and management. Yet, people are searching for answers. The global analgesics market is predicted to be worth more than $20 billion by the year 2020, and the pain management devices market over $300 billion (and that is just for physical pain; the antidepressants market is tipped to exceed $13 billion in 2018). These medical interventions are highly sought after because they can provide relief. What is often overlooked, however, is that medication itself cannot provide all the answers.

Another theme of the meeting that stood out for me was that pain is heavily grounded in our expectations. A great example of this is the rubber hand illusion. When people place their actual hand on the other side of a small divide so that it is out of sight, and a rubber hand is placed in front of them instead, this confuses the brain and people begin to respond to how the rubber hand is being treated. When people see it being brushed with a soft brush they report being able to feel the sensation, and when the experimenter suddenly pulls out a big wooden mallet and hits the rubber hand, people almost always jump as though they expected it to hurt. Professor Candy McCabe, who is the Florence Nightingale Foundation Chair in Clinical Nursing Practice Research at the University of the West of England, chaired a symposium titled 'Replacing Pain with Normal Perceptions

May be Therapeutically More Successful Than Trying to Remove Pain'. The talks in this session, along with many others at the meeting, suggested that pain is fundamentally based on our expectations, and correcting those expectations is an effective way to reduce pain. One talk, by Dr Jenny Lewis from the Royal National Hospital for Rheumatic Diseases, revealed how patients with chronic pain in their limbs experience visual illusions, perceiving those limbs to be larger and heavier, whereas in reality they are no different to normal ones. When these chronic pain patients put their hand under a screen on which the image of their hand could be reduced in size, thereby overriding their own misperceptions, their pain levels improved. The visual illusion of a smaller hand triggers the brain to think the hand is normal and this in turn reduces the experience of pain; it quite simply challenges the patient's expectations.

Perception is also fundamental to how we treat chronic depression, major trauma or other psychological disorders. These disturbances lead to changes in how people see the world around them, and when those perceptions are changed through psychological intervention, it reduces their symptoms. For instance, people who have had car accidents may experience feelings of anxiety or panic when they see the same model car that was involved in their accident on the road. Exposure therapy allows these patients to re-experience traumatic reminders while maintaining a calm mood, effectively pairing the reminder with a different set of expectations. The same basic principle is evident in treatments for depression. Depressed people tend to see their failures as diagnostic of their personal abilities or character, while their successes are passed off as mere chance. Changing how people interpret these events – sometimes viewing failure as a result of uncontrollable factors and success as reflective of their own personal qualities – helps to alleviate excessive negative self-perception and, in turn, improve feelings of depression. Rather than expecting failure, these individuals may sometimes expect success.

This is not to say there is no reality to physical or emotional pain. Those who experience terminal illness or the death of a loved one have very concrete reasons for their pain, and simply looking at things differently is not going to eradicate their suffering. There is very little that is positive about these experiences at all. Yet, we do know that how we respond to pain, how we make sense of it, wields a strong influence over our experience. From this perspective, being more fearless with pain and approaching it head-on, exploring its other qualities and developing insights into how and why in some contexts it may even have well-being benefits is certainly a worthwhile endeavour. There is no silver bullet, as I have said, so if nothing else such an approach allows us to take charge of our relationship to pain and our understanding of it, rather than allowing it to be cordoned off by medical diagnoses and clinical interventions.

WHEN IS PAIN A GOOD THING?

No matter what this book may have to offer those with chronic pain, trauma, major depression or extreme loss, the focus is mostly on the kinds of adverse events that can directly increase our happiness. So when is pain likely to be beneficial? What are the factors surrounding this experience that may help us differentiate good pain from bad?

At first blush the two obvious candidate factors are how intense our painful experiences are, and whether we feel we are in control of them. It is tempting to think it is just mild, as opposed to intense, aversive experiences that would have positive consequences and contribute to our happiness. It also seems logical that if we choose to endure a negative event, and therefore have control over it, our experience of that event should overall be more positive. The problem with relying on intensity and control as a litmus test of when pain is good as opposed to bad, is that people respond differently to their experiences, and some people are better able to cope in the same situation than others.

One way to understand this is to consider the personal resources that a person brings to an aversive experience. When people have sufficient resources to cope they will feel *challenged*, but when they have insufficient personal resources they will feel *threatened*. This means that different people may respond to the same painful event in very different ways. While the intensity of the experience and whether it is within the person's control play into this equation, they do not uniquely determine how people will respond. Some people are just more resilient than others.

Even here, however, whether people feel challenged or threatened may not be a good indicator of good vs. bad pain. As I noted above, sometimes people who are traumatized by adverse events grow from these experiences. This suggests that even intense, uncontrollable and threatening experiences can have positive outcomes.

Although intensity, control or the personal capacity to cope may go some way to telling us when we are likely to benefit from painful experiences in life, there are two key factors more important still. The first is whether our pain stops. Endless pain has never done anyone much good, and for this reason our negative experiences need a beginning and an end to be in direct service of our pursuit of happiness.

Luckily for us life is full of contrasts, allowing us to move between the experience of pleasure and pain. By understanding that it is exactly these contrasts which underpin our capacity for happiness, we can look for them more often, seek them out, and in that way find a more compelling, albeit more risky, version of happiness.

The second is how we choose to relate to pain. The modern-day message of acceptance is a half-hearted one. We are told if we accept something it will make it better, easier to bear. Yet, too often we keep one eye open to that possibility, hoping our discomforts will disappear. But is this really what it means to accept something, or to embrace it? We need bravery to accept things that hurt us, but when we see these experiences from a different perspective and understand that we need pain in more ways than we ever realized, we can learn to embrace our experiences more honestly and openly – and only then will true happiness come creeping up from behind.

PART ONE

Avoiding Pain

Charles Darwin struggled to comprehend why the world that he observed in his various travels, and the biological communities he studied, met with so much pain and misery. In his personal letters (*The Life and Letters of Charles Darwin*) he writes:

> But I own that I cannot see as plainly as others do, and as I should wish to do, evidence of design and beneficence on all sides of us. There seems to me too much misery in the world. I cannot persuade myself that a beneficent and omnipotent God would have designedly created the Ichneumonidæ with the express intention of their feeding within the living bodies of Caterpillars, or that a cat should play with mice . . . I feel most deeply that the whole subject is too profound for the human intellect. A dog might as well speculate on the mind of Newton. Let each man hope and believe what he can.

Darwin struggled with the very existence of suffering, yet he had a profound understanding of its value. At the very centre of his theory of evolution is the necessity of suffering, struggle and death; through these experiences, species can evolve and advance. This link between negative events and the production of beauty, wonder and advancement is evident in the following passage from *The Origin of Species*:

> Thus, from the war of nature, from famine and death, the most exalted object which we are capable of conceiving, namely, the production of the higher animals, directly follows. There is grandeur in this view of life, with its several powers, having been originally breathed into a few forms or into one; and that, whilst this planet has gone cycling on according to the fixed law of gravity, from so simple a beginning

endless forms most beautiful and most wonderful have been, and are being, evolved.

Like Darwin, there is a part of us that would prefer to see an end to suffering; to live in heaven or nirvana. Yet, by focusing our efforts on the attainment of utopia, we forget that it is precisely our hardships that allow us to grow, to expand our horizons, to become stronger and to experience the fullness of life. If we avoid all pain, we limit our capacity for growth and adaptive change and relegate ourselves to the pursuit of banal pleasure. If we fail to accept and value pain, we make it worse.

I

Have We Reached Peak Comfort?

> If tomorrow someone invented a foolproof, cost-free pill,
> with no side effects, guaranteeing life-time immunity from
> pain, we would at once have to set about reinventing what it
> means to be human.
>
> David B. Morris, *The Culture of Pain*

Never before in the world's history have as many of us had access to
as much income and resources as we do today. In 1820 there were only
a handful of countries that achieved any economic growth, and the
majority of the world lived in poverty with an income similar to that
of people living in the poorest countries in Africa today. By 1950 the
picture had changed dramatically, with people living in developed
countries approximately ten times richer than people living in developing countries. During the following three decades, however, this
pattern changed again, with developing nations (especially those in
South-East Asia) catching up those in the developed world. Now, relative to the 1970s, the world's average income has risen by approximately
400 per cent. While poverty remains a reality for millions, on average
the world has become much richer today than at any time in history.[1]

With more disposable income and greater access to resources, the
question arises: what are we spending all of these new-found riches
on? What kinds of things are both making us money and costing us
money? One way to understand this is to look at the most profitable
industries, those sectors receiving a larger share of our combined
wealth. According to *Inc.* magazine,[2] in 2014 the ten most profitable
industries were:

1. Health
2. IT Services
3. Business Products and Services
4. Energy
5. Financial Services
6. Human Resources
7. Logistics and Transportation
8. Consumer Products and Services
9. Construction
10. Telecommunications

It would appear we are spending more money on health services than any other economic sector – a fact that also corresponds with data showing the average life expectancy rising year-on-year. In 1800 there was no country on earth where the life expectancy was above forty years. Today even countries that are among the poorest in the world (e.g. Sierra Leone and Mozambique) have life expectancies over forty-five years, and the wealthiest countries have life expectancies over eighty years.[3] We are living twice as long as we used to two centuries ago and this is, in part, due to our economic largesse.

Although there seems to be a diversity of highly profitable industries in the list above, there are some basic similarities between them. While spending on health is clearly distinct from spending on financial services or telecommunications, if we break all this down to an analysis of *why* we spend money on these things one clear factor emerges: we are mostly seeking to maximize comfort. Health is a good example; our spending is focused on protecting our health so we do not experience pain and can delay death. Energy expenditure is largely motivated by our desire for heating, cooling, lighting and power for our electronic devices that make our lives easier or provide us with entertainment. In fact, about 30 per cent of our household energy consumption is allocated to the provision of hot water alone – which is in large part simply to ensure we have longer, warmer showers. We spend money on logistics and transportation to get us from A to B easily and efficiently, and without the inconvenience of walking long distances. We build homes and offices to ensure we are dry, warm and comfortable. We spend money on telecommunications which allow us to interact with others whenever we want and without the need to travel or wait. We spend money on consumer products such as beverages, food, toiletries and cosmetics to ensure that we are never hungry or thirsty and are well groomed.

Our ancestors busied themselves collecting wood for fires to cook and keep warm, hunted animals for food, and built shelters to keep dry all for the very same reasons. Deep within our psyche is a basic desire to make life easier, more comfortable and less physically and emotionally demanding. This is how we, and most other animals, have survived since the dawn of time. It is also what led the ancient Greek philosopher Plato to declare that pleasure is intrinsically good and pain is intrinsically bad.

Our pursuit of pleasure and avoidance of pain is rooted in the rational, functional and good. But it may just be we have overshot the mark. Whereas those who came before us sought to maximize their comfort, they were constantly surrounded by hardship and adversity. Collecting food was hard labour. They certainly did not walk down supermarket aisles trying to decide which type of milk to buy. They were exposed to the elements. They did not live in climate-controlled houses, sleep on rigorously tested luxury mattresses, drive in cars with heating, air-conditioning and adjustable seats, sit in offices with state-of-the-art ergonomic chairs, or walk in shoes with scientifically proven shock absorption. They experienced discomfort on a daily basis. They did not have a smorgasbord of medication available to them should they happen to experience a headache, period pain or fatigue. Today we have the capacity to achieve complete comfort in every moment of our waking and sleeping lives.

The trouble with this comfortable existence is that when pain does arrive it takes us by surprise and we find ourselves ill-equipped to cope. By this same measure, it is only pain that distinguishes itself when pitted against a baseline of complete luxury. We become numb to our 'simple' creature comforts, stripping them of their capacity to deliver any pleasure. This is why we are always on the lookout for ways in which to titillate our senses and tastes. More and more we seek out the sublime. Over the past twenty years there has been a sharp increase in luxury spending, with our worldwide total expenditure on luxury goods and services tripling.[4] We spend large amounts of money chasing beauty and youth. We also spend significant amounts of money on pleasant sensory experiences such as exquisite food and drink, the smell of exotic perfumes, and the restorative pleasure of health retreats, beauty spas and exotic resort-style holidays.

While a sole focus on increasing our comfort may have been rational and functional in times gone by, in our economically and technologically advanced societies where the bar is already set so high, such an approach may no longer hold the answer to our happiness.

BEYOND HEDONISM

In 300 BC another Greek philosopher, Epicurus, declared that pleasure is the only intrinsic value known to man and, therefore, everything is valuable only to the extent that it creates and enhances pleasure. Based on this theory, Epicurus developed the concept of *hedonism*, which dictates that one's goal in life should be to maximize pleasure and minimize pain.

We have, in many ways, moved closer to this goal than at any time in history. Of course, there have always been the lucky few (e.g., those with access to a wealth of natural resources, royalty, etc.) who have had the capacity to carve out a life full of enjoyment, yet today entire societies have reached a point where people are living in near-constant and relative comfort. Would Epicurus be proud, I wonder. He was no stranger to pain, having served in the military for two years. He would have walked long distances on a regular basis, battled the elements, and suffered from the pain of kidney stones, an affliction that eventually caused his death in 270 BC. He recounted his experience of this in a letter he penned to one of his students on his deathbed:

> I have written this letter to you on a happy day to me, which is also the last day of my life. For I have been attacked by a painful inability to urinate, and also dysentery, so violent that nothing can be added to the violence of my sufferings. But the cheerfulness of my mind, which comes from the recollection of all my philosophical contemplation, counterbalances all these afflictions.[5]

Epicurus may be the godfather of hedonism, but his life was certainly not devoid of challenges. He, less famously, argued against excessive pleasure, suggesting that overindulgence itself leads to unhappiness. So, this emphasis on pleasure was of a finite kind – enough pleasure as opposed to all pleasures all the time.

The link between an excessive ability to satisfy all our desires – to live in complete comfort – and feelings of unhappiness is often evident in celebrity culture. Take Christina Onassis, the daughter of shipping tycoon Aristotle Onassis. She inherited wealth beyond imagination and could afford a lifestyle that was barely conscionable. William Wright, the author of her biography, *All the Pain That Money Can Buy*, details how she would spend $30,000 to send a private jet to America to keep her stocked in Diet Coke, and once sent a helicopter from Austria to Switzerland to retrieve a David Bowie cassette she had left there. She was even reported to pay her 'friends' as much as $30,000 a month to ensure they had time to spend with her whenever she so desired. Dying from a heart attack at the age of thirty-seven, Christina was remembered for her struggles with her weight, her pill-popping, and her penchant for diamonds at breakfast.

The repercussions of excessive pleasure are not limited to celebrities or daughters of tycoons. The truth is, over-indulging in just about any form of pleasure will quickly have a downside. A friend of mine was recently introduced by a well-meaning colleague to the hugely popular dating app Tinder so that he might find a partner, or at the very least go on a few dates. At first, like many users, he was over the moon. As a child of the 1970s, the ease with which he could meet women using this new technology was astonishing. Some months later, however, he relayed to me with a hint of bashfulness how many dates he had been on, and, in turn, how many women he had slept with. In fact, he confided that he needed to take a break and it was getting totally out of control. Having willing dates on tap with not much more than a right swipe of his thumb had provided a valuable distraction for a time, but now it was causing other problems. The sudden assortment of sexual relations did not make him feel good and these short-term relationships lacked meaning. Beyond distracting him, these interactions were starting to erode his self-esteem, causing him to feel more alienated, depressed and lonely.

It would appear this effect of Tinder is not unique to my friend. In a study of 1,044 female and 273 male university students, researchers found that approximately 10 per cent used Tinder.[6] Being actively involved with Tinder, regardless of gender, was associated with dissatisfaction and feelings of shame about one's body, a tendency to monitor one's physical looks and to internalize societal expectations of beauty, and to compare oneself physically to others. People on Tinder often feel depersonalized and disposable in their social interactions. The world of fast hook-ups, it would seem, can provide unprecedented access to sexual pleasure, yet when human interactions are reduced to the satisfaction of basic desire, people quickly begin to feel like commodities – soulless and without substance. Most concerning is that many people, like my friend, find these kinds of interactions become addictive: they are always looking round the next corner in case something better comes their way.

When it comes to enjoyment, more is not more. If you use a small amount of a drug for the first time you will get high, but do it again and you will need more as your tolerance increases. Further, maintaining that high is near impossible. This is why drug addicts continue to use. They keep chasing the same high they got from the first hit, but each time they need a little more to get there; a behavioural pattern that has dire consequences. Pleasure dissipates over time, and the more pleasure we have the more advanced the experiences we need to obtain a comparable high.

I recall a time when I was on a long hike in Chile. It was about day five of a seven-day ordeal. We had been walking up and down steep mountainsides, and had endured a snow storm and recurring altitude sickness. As we came down the other side of the mountain range we came across a house where they sold beer. Even better, as we were buying ourselves a well-deserved beer the owner informed us that just a little down the path there was a nice spot to camp right next to a natural hot spring. I will never forget the absolute pleasure of sitting in those hot springs, resting our weary feet and drinking that beer. The fact is, the beer was actually warm, and the hot spring stank of sulphur, but at that moment it was heaven. What made this so pleasurable was not the quality of the hot spring or the beer, it was the five days of hiking, the cold, the snow and the sickness that came before it. We left the next day, but I am sure had we stayed there for another day or two I would have begun to notice the poor quality of the beer and the smelly water of the hot spring. The pleasure would have quickly dissipated.

Compare this experience to what we might readily reel off if asked to detail our ideal holiday. It would probably involve golden beaches, luxurious accommodation, endless gourmet food and drink, and possibly a health spa. This would be wonderful, but would it be more wonderful than a smelly sulphur puddle and a warm past-its-sell-by-date beer when consumed after five days of eating not much more than pot noodles and instant mashed potato? What if you stayed there all the time, just living the dream as many of the rich and famous do? Would you wake up each day absorbed in the pleasure of it all, or would it eventually become a little mundane? The truth is that if you wanted to experience that same 'high' you got the first moment you arrived, you would need either to go for a five-day hike or upgrade to an even fancier hotel. Our misguided approach to pleasure is evident when we take the second option and try to maximize pleasure while also avoiding all pain. We chase more and more pleasure – better food, better cars, better houses, better clothes, better holidays – just so we can experience any pleasure at all. The trouble is that this all works very much according to the law of diminishing returns.

While most of us do not live the life of Christina Onassis, we have an unprecedented capacity to achieve a certain hedonistic ideal. Compared to our early ancestors, for whom pleasure was a rare joy, and hardship and pain were simply part of daily existence, we are now confronted with a new set of problems. Specifically, how can we continue to achieve pleasure and happiness in a world relatively free of pain?

THE HEDONIST'S PARADOX

The basic cultural impulse to maximize pleasure and reduce pain is not only evident in consumerism, but is implicit within many forms of religion, underpins most medical or psychological interventions, and characterizes the received wisdom of the modern-day self-help movement. So pervasive is this assumption, we rarely stop to consider what it would mean if we achieved this goal with absolute impunity.

David Pearce is a British philosopher who believes we not only should, but could, work towards the abolition of all suffering. In 'The Hedonistic Imperative',[7] an extended online essay, he outlines a vision for the end of all suffering through the advanced application of technologies such as genetic engineering, nanotechnology, pharmacology and neurosurgery. He believes that these technological advancements will converge to end all forms of suffering among human and non-human animals alike – a project he refers to as 'paradise engineering'. So strong is his conviction that, on his website, he predicts 'the world's last unpleasant experience will be a precisely dateable event'. In a very Huxleyesque kind of way, he suggests that just as anaesthesia has ended physical pain, soon we will have the ability to end mental pain. There are four key reasons why I think we should be sceptical of Pearce's project.

1. Humans adapt to their environment

One reason endless pleasure is unlikely to be idyllic is because we have an inbuilt ability to adapt to our environments. Through the process of habituation, both pleasure and pain become less intense over time. Whether it is the reward of a massage or the sting of icy-cold sea water, over time (and within limits) we can get used to these experiences. This has allowed our ancestors to be able to adapt emotionally to living in harsh and unwelcoming environments.

Our capacity to adapt is also the reason why spending the rest of our lives in a luxury resort would, over time, fail to provide a great deal of pleasure. It might at first, but it would soon dissipate. Just as jumping into a hot Jacuzzi provides instant pleasure, over time our enjoyment is reduced. We adapt to the heat of the spa, and the only way to get that pleasurable sensation all over again is to exploit the contrasting experience of jumping into a cold pool.

2. Hedonic experiences are relative

How would we ever know what pleasure is if we experienced nothing else? Our hedonic experiences are relative (something that we will return to in Chapter 3), and pleasure and pain are only understood in relation to their opposites. Whether it be sitting in a Jacuzzi, lying on a beach, watching a beautiful sunset, enjoying time with friends . . . the list goes on, all of these experiences lose any sense of pleasure if they are not contrasted with something else, most commonly work of some kind. It is the strength of the contrast that makes all the difference. Jumping from a lukewarm bath into a hot spa is not as amazing as jumping from a freezing cold pool into a spa. Just as seeing friends after a period of isolation is always more enjoyable. Yet, even here the contrasts can be strengthened: finding someone you thought you had lost, or enjoying a date with your partner for the first time all *year* because you had been overwhelmed with childcare duties. Our unpleasant experiences provide an important avenue for pleasure and the sharper the hedonic contrast the more pleasure we feel.

Another reason why highs as well as their corresponding lows are so critical for our overall happiness has to do with how we construct our experiences retrospectively. Our global evaluations of events tend to be heavily swayed by the high or low points as well as the end points. This is what Barbara Fredrickson refers to as the peak-and-end rule.[8] Together with Daniel Kahneman, Fredrickson tested this idea in a study where the researchers asked volunteers to view a series of twelve video clips.[9] These clips varied in two ways: they were either pleasant (e.g. of ocean waves) or unpleasant (e.g. of corpses) and lasted for either 30 seconds or 90 seconds. As they watched the clips the volunteers provided moment-by-moment ratings on a sliding meter that allowed them to indicate how positive or negative the clip made them feel. Afterwards they provided global ratings of how positive or negative the clip was overall. What the researchers found was that participants' global evaluations of each clip were best predicted by their peak ratings while watching the video clip as well as their ratings right at the end. It was not the overall length of the clips that mattered. As the researchers suggest, memory does not take film, it takes photographs, and these photographs are taken at predictable points – peaks and ends.

Fredrickson suggests that we keep track of, and remember, the peaks and troughs of our experiences because they provide two important pieces of information. The first is about the experience itself. It is only by remembering the worst moments of a relationship break-up that we can really know how bad such an experience can get. The second is about our personal capacities to cope. It is only at the extremes (the peaks and troughs) that our personal resources are tested to their limits, and as such it is these moments that show us whether we have the personal capacities for achieving, enduring or coping with a specific experience. This is true of negative as well as positive experiences. Just as an intensely negative experience may require certain character strengths to endure it, intense positive experiences do also. As Fredrickson points out, 'just as you need to know the maximum height of the sailing boat you are towing before you drive under a low bridge, peak affect is worth knowing to decide whether you can handle experiencing a particular affective episode again'.

Returning to the principle of relativity, we can now see that the shift from pain to pleasure serves to make us happy in two key ways. First, painful experiences increase the intensity of subsequent pleasurable experiences. Secondly, this relative contrast means that we experience more intense peaks as well as more intense lows, and both contribute to a sense of personal meaning in our lives. It is for this reason many of the experiences in life that produce happiness and fulfilment involve both pleasure and pain. From this perspective, it is fair to say that skydiving is likely to contribute more to your overall sense of happiness than sitting at home watching television. The prospect of jumping out of a plane fills us with both excitement and dread simultaneously and this makes it especially pleasurable, meaningful and memorable. If it were not for our innate fear of falling thousands of metres through the air, skydiving would not be especially enjoyable. It also would not provide much insight into the limits of our personal resources, and therefore would not be especially memorable. Critically, however, we recall the act of skydiving as fun, enjoyable and something that made us happy. Our memory tends to focus more on the excitement than the dread.

3. Pain and pleasure are two sides of the same coin

Another reason to be sceptical of David Pearce's project is that pain and pleasure are linked at the physiological level (another point covered in detail in Chapter 3) and, therefore, numbing our pains may have the side effect of numbing our pleasures. This was demonstrated in an experiment by a group of researchers from Ohio State University.[10] Across two studies, 167 undergraduate volunteers were either given 1,000 mg of the painkiller acetaminophen (paracetamol) or a placebo. After waiting an hour for the medication to begin working, the experimenters showed the volunteers forty pictures, ranging from extremely unpleasant to extremely pleasant. Their task was to rate the pictures on a scale from -5 (extremely negative) to +5 (extremely positive). They then viewed the same forty pictures again (in a different order) and were asked to rate how emotionally aroused they felt when viewing each image, on a scale from 0 (I feel little to no emotion) to 10 (I feel an extreme amount of emotion). In one of the studies, the experimenters also had the volunteers rate the pictures a third time on the extent to which the colour blue was present in the image, using a scale from 0 (the picture has zero colour blue) to 10 (the picture is 100 per cent the colour blue).

Results showed that overall the volunteers who had taken the painkillers were less extreme in their emotional response to the pictures than participants who had taken the placebo, and this was the case for both positive and negative images. In both cases these differences were more pronounced for the more extreme stimuli – extremely positive and extremely negative pictures both had less of an impact on volunteers who had taken the acetaminophen. The researchers noted these differences were not apparent for judgements of colour, indicating that the painkillers specifically reduced the positivity and negativity of the pictures (their pleasantness and unpleasantness) and emotional reactions to their content, but did not affect perception of other elements.

This goes to show that when we reduce our experiences of pain we are just as likely to reduce our experiences of pleasure. Killing pain narrows our emotional bandwidth in life. We experience less intensity, less emotion, and therefore our 'hedonic tone' becomes restricted and less variable. We experience fewer troughs, but we also experience fewer peaks.

4. Our inner experiences are paradoxical

Another reason to be sceptical that we can achieve endless pleasure has to do with the paradox of hedonism. This concept was first explicitly noted by the philosopher Henry Sidgwick in his work *The Methods of Ethics*. The basic idea is that true happiness cannot be acquired directly, it can only be acquired indirectly. In this way, happiness (like many of our emotional states) does not appear to operate in the same way as other things in the world because you cannot directly seek happiness like you can seek riches.

One way to make sense of this is through understanding the psychology of goal pursuit. When we set a goal for ourselves we tend to assess our progress towards that goal. Because important goals take time to achieve, there will be many occasions when, upon reflection, we realize we have not yet attained all that we want to. It is at these moments we feel a sense of disappointment, but this sense of disappointment can be valuable as it serves to motivate us to work harder towards attaining our goal. The paradox of hedonism arises when we set emotion goals such as the ideal of feeling happy. This was illustrated by Jonathan Schooler from the University of California in a 2003 paper.[11] He and his colleagues noted that when we reflect on our progress towards important emotion goals, the feelings of disappointment we experience along the way play into our goal attainment differently. Rather than motivating us to work harder to achieve our goal of happiness, our feelings of disappointment directly interrupt our efforts. Rather than feeling happier than we were before, we now feel disappointed – an emotion state that certainly does not contribute to our overall levels of happiness and may even detract from them. This led Schooler and his colleagues to suggest that making happiness a goal in life is in fact counterproductive. Trying to be happy may make us less happy than we were before.

According to Pearce we should aim to seek pleasure, yet as the paradox of hedonism illustrates, the key problem in trying to achieve pleasure is that it tends to evade us every time we do. Like a shy creature residing in the recesses of our mind, pleasure does not take kindly to being in the spotlight. Pleasure can only be achieved indirectly by pursuing other things in life, not because they will make us happy, but perhaps because they are meaningful and important in other ways. When we make pleasure itself the end goal, we are likely to experience less pleasure than we might otherwise.

LOVING PLEASURE AMPLIFIES PAIN

The other problem with hedonism is that by placing a premium on pleasure, it also serves to devalue pain. This philosophy tells us that if we experience pain we are failing to achieve pleasure, and if maximizing pleasure should be our life goal, then we are failing to live a *worthwhile* life. When pain is not only a signal that we are hurting, but a signal that we are *failing writ large*, our experience of it is intensified.

A powerful example of how our understanding of pain can fundamentally change our experience of it was noted by Henry Beecher during the Second World War. A professor of anaesthesia at Harvard Medical School, Beecher served in the US Army in North Africa and Italy. In a 1946 paper he noted that despite the common belief that flesh wounds are inevitably associated with pain, observation of recently wounded men in the combat zone showed this generalization to be misleading.[12] Beecher found that among seriously wounded soldiers who were questioned within twelve hours of receiving their wounds, 25 per cent reported only slight pain, and 32 per cent reported no pain at all. Furthermore, 75 per cent did not want any medication for their pain. This surprised Beecher and also led him to try and understand why it could be that wounds, which in these cases were severe and would normally be very painful, seemed to produce almost no pain at all. He determined that the soldiers were on balance happy to be wounded. It meant that they could go home, the momentous prospect of which alleviated their physical ailments. In the context of the war their wounds literally represented safety from possible death. In Beecher's words: 'his wound suddenly releases him from an exceedingly dangerous environment, one filled with fatigue, discomfort, anxiety, fear, and real danger of death, and gives him a ticket to the safety of the hospital'. For these soldiers, the pain itself was a relief.

Beecher's observations underline the problem with devaluing pain, or framing it as simply something that is 'bad'. More recent research shows that how we frame pain can change how we respond to it, even at the neurological level. Fabrizio Benedetti from the University of Turin Medical School in Italy exposed two groups of volunteers to the experience of pain by restricting the blood flow in their arm while they squeezed a hand spring exerciser (this kind of pain can become intolerable within about 14 minutes).[13] One of the groups of volunteers received standard instructions for the test, which simply stated the experience might be painful. The other group of volunteers, however, were told that enduring the pain would be beneficial for them – specifically, it would enhance their muscle performance. So, one group were given the standard negative framing ('this will hurt'); the other group were given a positive framing ('this will build muscle').

Researchers found that the participants who were told the pain would have positive effects were able to tolerate more pain compared to the participants given the standard instructions. Researchers were then able to show that these effects were due to the increased activation of neurotransmitters that commonly help people cope with pain (opioids and cannabinoids). Simply changing the purpose of pain led the brain to alter how it processed the message of pain.

A particularly poignant illustration of just how much our experience of pain is based on our expectations comes from work on the placebo effect. Placebos are used in clinical trials when testing the effectiveness of a new drug. Patients will be split into two groups, with one group receiving the actual drug and the other receiving a sugar pill. These trials are generally called 'double-blind trials' as both the patients and researchers do not know which group is the placebo group and which is the experimental group until the end of the study, thereby reducing the possibility of bias. Researchers in these trials began to notice that patients in the placebo groups would often show significant improvement in their symptoms. Of course, this was inconvenient for the researchers as it meant that for any new drug to be deemed 'effective' it needed to outperform the beneficial effects of the placebo.

In order to better understand the placebo effect, researchers have begun to study it on its own terms. In one study, researchers from Oxford University administered a potent opioid analgesic (remifentanil) to volunteers while they were exposed to thermal pain (heat).[14] This took place in a functional magnetic resonance imaging (fMRI) scanner so the researchers could examine neural responses while the volunteers rated the intensity of the pain and their feelings of anxiety with regard to it. At the beginning of the study, the experimenters first exposed the volunteers to the thermal pain while they delivered a saline solution through a needle in their arm. This provided a 'baseline condition' to measure against. On the next trial volunteers were told that they would be administered an opioid analgesic, when in fact they were still receiving the saline solution. This was the 'no expectation condition'. In the next trial, they were administered the opioid analgesic, but this time they were told they were being given a particularly *effective* drug. This was the 'positive expectation condition'. On the last trial they were given the same analgesic, but told they were being administered a particularly *ineffective* drug. This was the 'negative expectation condition'.

Results showed that while the opioid analgesic reduced pain even when participants had no expectations of its effectiveness (that is when they instead thought they were receiving a saline solution), its analgesic benefits doubled when they expected it to be effective. Just so, when they expected it to be ineffective, the analgesic benefits of the drug were completely abolished; it was no more effective than the saline solution when they were expecting just that.

The fact that our expectations can so fundamentally alter the effectiveness of a drug provides a major challenge to the pharmaceutical industry. Drugs are rolled out on to the market once they have been tested and shown to be effective but not harmful. This means they have been demonstrated to be *more* effective than a control baseline condition where people expected the benefits of a drug but receive a placebo instead (usually saline solution or a sugar tablet). Yet, it is important to note that this placebo effect is equal across both conditions. Whether people are taking the actual drug or a fake drug, they are still experiencing the psychological effect of pill taking, and this makes them feel better independently of any actual pharmacological effects. This means that when we pop a Panadol or an aspirin we are not only getting the benefits of the medication, we are also getting the psychological benefit of the expectation that taking the pill will make our pain better. Shifting our expectation can almost totally undermine the effects of the medication, and our expectations can sometimes have a stronger effect than the drug itself.

This further suggests that the packaging for medication may be just as important as the medication itself, and this is not something that has been lost on the pharmaceutical companies. Take for instance a court case won by the Australian Competition and Consumer Commission against Nurofen, a leading analgesic brand in the country.[15] The company had been selling so-called specialized pain relief products for almost double the price of their standard ibuprofen product. These 'targeted' analgesics claimed to be able to specifically treat migraine pain, tension headaches, period pain and back pain – with a separate product for each pain type. The trouble was that each of these products contained exactly the same ingredients as the standard brand. The only benefit consumers were getting from these more expensive versions was different packaging, and the belief their pain was being directly targeted. Although flirting with consumer fraud, it worked quite effectively.

The importance of packaging on the effectiveness of analgesic medication was confirmed for me in my discussions with an anaesthetist friend who told me that he regularly employed the placebo effect in his pain treatment clinic. He noted it would be relatively easy for him to just jab a needle into his patients and send them on their way, but he felt this approach was less effective. Rather, he would bring his patients into the clinic, ask them about their pain, and then inform them they would need to come back for an injection the next day. He would let them know the injection would hurt, the needle was rather big and as such it would cause significant pain, but there would be a kind nurse present who would look after them. This anaesthetist believed that by setting up the administration of each injection as a significant event, this led patients to expect more from their treatment, and this in turn improved their outcomes.

Telling people an injection might be especially traumatic may sound counterintuitive, and perhaps even a little sadistic, yet research has demonstrated that more painful (and therefore more significant) placebos tend to work better. In one trial researchers compared the effects of using a sham acupuncture needle to taking a placebo pill in terms of pain relief.[16] In both cases the treatments themselves had no benefit beyond the placebo effects that they triggered. In this study the needle produced more positive effects, leading to a greater reduction in self-reported pain and reduced severity of pain-related symptoms. Needles are more significant because they hurt more, and we therefore expect they are going to have more powerful effects – an expectation which ensures that they do.

FEARING PAIN MAKES IT WORSE

There is now a long line of research which shows that fearing the experience of pain can make it worse. In a study of 104 chronic pain patients researchers found that the best predictor of disability was not the intensity of the pain these volunteers experienced, but rather their dread.[17] The more that these patients tried to avoid their discomfort, the worse their level of disability. Beyond increasing disability, it is also well-established that a fear of pain leads to an increase in our experience of it. In trying to understand why some patients tend to develop more pain-related problems than others, researchers have suggested avoidance theoretically leads to exaggeration,[18] while confronting and accepting physical complaints should lead to contextualization and reduction.

This logic underpins our understanding of panic disorder. One reason why people have repeated panic attacks is their *fear* of panic attacks – that is, their feeling of wanting to run away from the panic itself. The experience of a sudden onset of panic can occur for a variety of reasons, and it triggers a 'fight or flight' response, which is functional and valuable; this is exactly what you want to experience when faced with a hungry lion. What perpetuates the experience of panic, however, is that people become fearful of the experience of panic itself. Often they interpret the experience as an indication that they are losing their minds, or perhaps as a sign that there is something wrong with their heart (which generally feels like it is beating out of their chest). When a panic attack occurs, it is usually out of context, and there is no hungry lion to help make sense of the experience. Rather than representing the presence of an actual threat or physical illness, a panic attack is caused by the misfiring of the flight–fight system – a completely safe but misplaced physiological response. When this happens for no apparent reason it is bound to unsettle. Going forward, people can become sensitized to an increased heartbeat, sweaty palms or feelings of light-headedness, all of which are interpreted as an indication they are about to have another panic attack. This all but guarantees that panic sets in – people panic about having a panic attack, and the cycle continues.

There are a variety of treatments for panic disorder, but in my experience one of the best is to tell people to *try* and have a panic attack. I often tell my patients that the next time they feel their heart racing, or experience a little light-headedness, they should just take a moment and try as hard as they can to panic about it. The fact is, if done in earnest, this can almost entirely prevent the attack. Just as running away from something that feels threatening increases the sense of threat, running towards something that feels threatening (when it is not actually threatening) does the opposite – it reduces our emotional response.

This same pattern was observed in a study by Eddie and Cindy Harmon-Jones, now at the University of New South Wales, where they asked volunteers a number of questions about their attitude to a range of emotions.[19] For instance, in the case of fear the volunteers indicated on a scale from 1 (rarely/never) to 5 (almost always/always) whether 'I like to do things that scare me' or 'I dislike doing things that scare me'. In the case of disgust, they indicated whether 'If I smell something disgusting, I will smell it again on purpose' or 'I do not enjoy doing things that I find disgusting'. The items about liking were reverse scored and added to the items about disliking in order to create an index of how much people disliked each emotion. They found that the more people dislike emotions such as fear and disgust, the more frequently they tend to experience them. Like pain, these are emotions that serve to warn us of threat (in the case of disgust it is the threat of disease), and therefore disliking them or trying to avoid them increases their frequency and intensity. In contrast, people who disliked these experiences less – those who naturally tried to avoid them less – also experienced them less.

It was exactly this insight which led Jon Kabat-Zinn, professor emeritus of medicine at the University of Massachusetts Medical School and author of *Full Catastrophe Living* (and commonly regarded as the father of the current mindfulness movement), to use mindfulness meditation as an intervention with his chronic pain patients. He realized that they had been mentally trying to avoid their pain, running away from it and simplistically labelling it as bad. His approach was to ask them to focus their minds on their pain, to understand where it was and what it felt like, and to avoid evaluative labels such as 'it feels bad'. In Kabat-Zinn's words: 'You change your relationship to the pain by opening up to it and paying attention to it. You "put out the welcome mat" not because you're masochistic, but because the pain is there', demanding to be heard.[20]

Seeking to avoid unpleasant experiences only tends to make them worse. Rather, confronting, accepting and even approaching them is a better management strategy. Fortunately, we have good reason to approach pain on the front foot; such experiences are in fact critical for our happiness. A life without any discomfort at all would be quite a horrible existence. This realization motivated Aldous Huxley to write his book *Brave New World* in 1931. He imagined a world in which people had the capacity to eradicate their negative experiences by taking 'soma' (a substance that is eerily similar to modern-day antidepressant medication). Those living in this brave new world could eradicate all their discomfort, yet it was exactly this capacity that Huxley believed made the society he wrote about *dystopian*. He believed such an existence would have been freakishly banal and bereft of any real meaning. A life free of any suffering at all would be without distinction, without edges, and without variation. A world without pain would be torturous indeed.

If you stop to consider the happiest moments in your life, they are usually experienced on a knife-edge between pleasure and pain. Whether it be finding your true love, holding your newborn baby for the first time, or a great professional achievement, all of these moments of happiness are couched in the potential for suffering, loss or failure. Love is a powerful experience because it makes us vulnerable; grand achievements are only grand because those who achieve them risk humiliation; and holding your newborn in your arms for the first time occurs in the context of great insecurity, anxiety and pain. If we eradicate the possibility of pain, loss or failure, we eradicate joy and strip these experiences of any kind of lasting meaning.

2

The Cotton Wool Generation

The Australian history of my family probably began in Victoria, where they arrived on ships from Britain. From here, they headed .west for the gold rush around 1900. They made their way to a small town called Day Dawn, which is about 650 kilometres north-east of Perth and in the middle of nowhere. It was built around a large mine that employed approximately 750 people, which when it closed down took the town with it. The Bastians took the train as far as Mullewa, where the line ended, and then walked the final 300 kilometres alongside a camel train. On arrival, they had to build somewhere to live and establish the small business that eventually funded their livelihood.

That was where my great-grandfather grew up, no more than a young boy at the time of their epic journey. Later, his son, my grandfather, became a sheep shearer. He would spend long periods of time moving across the desolate backcountry of Western Australia to various farms to find work. Summer days would have been sweltering, with little respite from the heat. Occasionally, he would have showered with a bucket and a cloth.

For my generation, this hard lifestyle is a distant memory. I sit writing these thoughts in my perfectly cooled office, on my ergonomic office chair with sophisticated comfort adjustment, considering what I should have for lunch – sushi, salad or falafel? It's a world of occasional hard choices, but a world pretty much divested of true hardship. While there have always been those whose wealth and privilege have protected them from the need to be tough, the relative comfort and safety of today have, over the last few decades, entrenched themselves as the norm for a great many more of us. For better and for worse, we no longer need to be physically tough like those who lived a generation or two ago did.

However, there are plenty who still struggle to get by with very little respite in their lives. These people endure a lifestyle that makes the experience of Australian settlers look like a walk in the park. In many places around the world, whether in the slums of cities or in sparsely populated villages, large segments of the population still sleep on dirt floors, wash with buckets of cold water (even in winter), which need to be hauled from communal taps, and walk long distances on a daily basis. They need to boil their water, often on coal fires, before they can drink it, and live mostly on a restricted diet of cheap carbohydrates. Sickness and infection are common, with little or no relief provided by modern medicine. When it rains, they live with the mud and when it doesn't, they live with the dust.

When we consider these living conditions from our first-world position, they seem vastly different. They are also likely to be outside the parameters within which most of us could cope. While I am not suggesting such an existence is desirable, it provides for an important reflection on how our comfortable lives are shaping us and our children. As a member of what I will call 'the cotton wool generation', I have found it is not only a case of enjoying our comforts; we become infuriated when they are taken away. We expect a comfortable lifestyle, and we believe we deserve it. One minor hardship – a broken iPhone or mobile service disruption, a towed car, a delayed train – is enough to send us into a frenzy, as though we must be entitled to compensation, or at the very least some kind of recognition.

BENIGN MARTYRDOM

A few years ago I conducted a study at the University of Queensland with my colleague Jolanda Jetten.[1] We had become interested in the notion that people may seek to punish themselves when reminded of something immoral they have done in their past. In the psychology community, researchers mostly believed that self-punishment did not exist, as it was hard to find a motivation for such behaviour. Why would people *choose* to punish themselves? It seemed there were few benefits to enforcing pain and punishment on oneself. Nonetheless, this lack of a psychological motivation appeared to be at odds with the evidence of self-punishment within many religious traditions, where this kind of behaviour synthesized a form of repentance. Such acts are evident in the case of Shia Muslims who whip themselves with *zangirs*, whips made of knife blades, until their backs are covered with blood. In the Hindu *kavadi* ritual, participants use meat hooks and skewers to pierce their legs, face and tongue. Over the years, Christians have engaged in various forms of self-punishment, ranging from wearing hair shirts and chains to various forms of self-flagellation, even self-castration.

This led us to consider whether these examples were part of a general response to feeling shame or guilt.

In the experiment, we asked people to put their hands into a bucket of ice-water for as long as they could tolerate. A bucket of ice-water is perfect for inducing pain in controlled settings, as it will not cause any lasting pain or injury but is uncomfortable enough. The question was, would people hold their hands in the ice-water longer when they were made to recall an immoral deed, and, if so, would the ice-water also serve to make them feel better about what they had done?

We asked our volunteers to come into the testing room one by one. We asked some of them to spend five minutes writing about a time when they had behaved unethically, and the others we asked to write about an everyday experience. After this, we gave them all a list of words describing different emotional states and asked them to indicate how strongly they were feeling each (from 1, not at all, to 5, a lot). The word 'guilty' was embedded among a number of other words with equal emphasis to the rest. We had also framed the essay task as a memory test.

Next, we told people they were going to take part in a second study that was unrelated to the first, and that it was a test of their physical ability. Some of our volunteers were given a bucket full of ice-water and told that their task was to hold their hand in the ice-water, with their wrist submerged, for as long as they could. We also used a control group here, who were given the less painful task of holding their hand in a bucket of room-temperature water until we told them to stop. We then asked everyone to indicate their emotions again with the same list as in the first part of the study, and asked them to rate how uncomfortable the ice-water had been on a scale from 1 (no pain) to 5 (severe pain).

The results were very satisfying. First we found that, when participants had recalled an unethical deed, they held their hand in the ice-water for longer (an average time of 87 seconds) compared to those who recalled an everyday experience (an average of 64 seconds), an indication that people *had* sought to punish themselves. But could it be that feelings of guilt had swamped the physical discomfort among those who recalled an unethical deed? In fact, we found the guilty participants had rated the ice-water as more painful than the non-guilty group: their average pain score was 2.79, compared to the non-guilty average of 1.91. This could partly be because they held their hands under for longer, but it proved not only had they felt the pain, but they sought it out.

Perhaps most revealing of all were results directly pertaining to guilt. As expected, participants who wrote essays about a moral transgression rated their guilt higher than the control group. However, after completing the ice-bucket task, the 'guilty' participants rated their level of guilt at 1.11, about the same as those participants who had never been made to feel guilty in the first place. Those who had written about an unethical deed, but had their hands in room-temperature water, rated their guilt at 1.53, the highest of all the groups.

This study received a lot of media attention when it was published. One reason for this was it was released around the same time as the Christian observance of Lent, a period during which Christians intentionally engage in acts of fasting and self-denial in order to cleanse them of their sins. The notion that pain could resolve guilt resonated with this yearly practice.

Fascinated by the idea that people respond to their experiences of pain as punishment, we wondered: if pain could resolve feelings of guilt for the sinners, what would it do for the saints?

We set up the experiment again.[2] This time, however, prior to completing the physical tasks we had some participants write an essay about a time they had performed a good deed (our saints), while others were asked to write about a time they had done something unethical (our sinners). After they had written the essay and completed the ice-water task, the researcher told them they had to leave the room to get some materials from the photocopier for the rest of the study. Before they left, the researcher pulled out a bowl of sweets. They told each volunteer that they were left over from a previous study, and to feel free to help themselves. Left with a large bowl of sweets in front of them, and no one watching their behaviour, volunteers took as many of the sweets as they wanted. Of course, we were watching their behaviour: we knew that there were exactly seventy-five sweets. Once participants had left the room, the experimenter counted the number of sweets left in the bowl.

Those who experienced the ice-water did, as anticipated, take more sweets. But only the saints among them. When people were made to think of themselves as virtuous, and then had an experience of 'unjust' pain, they felt they deserved a little more from life (they took an average of 5.29 sweets from the bowl). Those who wrote about an ethical deed but *did not* experience any pain took 3.06 sweets, and people who wrote about an unethical deed (the sinners) and *did* experience pain took 3.05 sweets. People who wrote about an ethical deed and did not experience any pain took about the same number of sweets as people who wrote about an unethical deed and experienced pain. So it was only those who wrote about an ethical experience and experienced pain – the unjustly punished saints – who took markedly more.

Confident that people interpret pain as punishment, leading to a tendency to self-reward when that punishment feels unjust, we decided to ask one last question. If pain resolves guilt, and increases self-reward, would it be especially likely to lead people to indulge themselves in guilty pleasures?

In our next study we replicated the same ice-bucket procedure as before, but without asking people to write anything. We already know that people tend to focus on their better, more virtuous qualities when left to their own devices, so we could assume that people tended to see themselves as relatively moral even without writing an essay. After the physical task, the experimenter again left the room after excusing themselves, leaving a bowl of things 'left over from a previous study' for the participants to indulge themselves in. This time, however, people were told they could take one item from the bowl, which was filled with ten Caramello Koalas and ten highlighter pens. For anyone who has never been fortunate enough to sample the Caramello Koala – a staple of Australian traditional confectionary – they are a koala bear-shaped chocolate filled with caramel (the clue is in the name).

Using the same approach as before, we counted the remaining items, and found that people who had experienced pain chose the chocolate 60 per cent of the time, while people who did not experience any pain chose the chocolate only 26 per cent of the time. Pain led people to indulge themselves in pleasures that would otherwise make them feel guilty. We knew this because prior to running the study we had asked a smaller group of volunteers to rate the highlighter and Caramello Koala on a number of qualities. This confirmed that while people rated both as having similar economic value, they said they would feel more guilty if they chose the chocolate compared to the highlighter.

Having found that 'sinners' feel less guilty after pain and 'saints' feel more deserving, there was still one more small piece of this puzzle to resolve. Before participants did any of these tasks we asked them to complete a survey where they rated the following questions on a scale from 1 (strongly disagree) to 6 (strongly agree):

1. It bothers me when others receive something that ought to be mine.
2. It makes me angry when others receive an award which I have earned.
3. I can't easily bear it when others profit unilaterally from me.
4. I can't forget for a long time when I have to fix others' carelessness.
5. It gets me down when I get fewer opportunities than others to develop my skills.
6. It makes me angry when others are undeservingly better off than me.
7. It worries me when I have to work hard for things that come easily to others.
8. I ruminate for a long time when other people are being treated better than me.

9. It burdens me to be criticized for things that are being over-
    looked with others.
10. It makes me angry when I am treated worse than others.

Together, these ten questions are designed to measure 'personal justice sensitivity'. That is, the extent to which people are especially
concerned about whether or not they are treated fairly and get what
they deserve in life. What we found was that when we looked at
people who tended to agree with these items, they were the ones especially likely to take the chocolate after experiencing pain. People who
tended to disagree with these items were no more likely to take the
chocolate whether they had experienced pain or not.

It drove home this notion that people subconsciously correlate
pain with some kind of punishment. The physical experience of discomfort activates justice-related thinking, meaning people feel less guilty
and more deserving after painful experiences. These findings point to
one of the benefits of engaging· in uncomfortable experiences and
helps to explain why people, when they feel guilty, often go out for a
hard run or a tough workout at the gym, or perhaps may engage in
other less beneficial forms of pain, such as in the case of those who
self-harm, because these experiences make them feel like justice has
been served. It also highlights why we are perhaps likely to reward
ourselves after enduring a challenging experience – we feel more
deserving of rewards because they serve to rebalance the scales of
justice so to speak.

Beyond these short-term psychological benefits, the results also
reveal that for those who score high on justice sensitivity, painful
experiences appear to run more starkly against their expectations of
how life should be. Simply experiencing the cold of an ice-bucket
was sufficient to make these individuals feel entitled to something in
return.

In many ways our lives have become so full of creature comforts,
and so free from pain, that for some at least painful experiences trigger a sense of indignation and, in turn, entitlement. It makes me
wonder whether those earlier settlers in Australia or others around
the world who endure a range of adverse and strenuous experiences
daily would respond in the same way. Have we become so used to our

comfortable lifestyles that simply placing our hand in a bucket of ice-water has a significant and measureable psychological impact?

SUFFOCATED BY COTTON WOOL

The emergence of this cotton wool culture – the personal conviction that our lives should be free of hardship – is linked both to our parenting practices and our educational systems. One reason to look towards parenting and education is that increasingly those spheres promote an 'everyone wins' mentality. This is part of a much broader phenomenon that can be traced back to the era of the baby boomers. Then, young and full of ambition, the generation born in the aftermath of the Second World War saw their potential as unlimited, and in some ways they were right: the population spike meant they held demographic sway, and the decades after the war brought an era of unprecedented economic prosperity. Throwing off the shackles of conformity and conservatism, these baby boomers believed no one should tell them they couldn't have their cake and eat it too. Just as the lyrics to the hugely popular Hot Chocolate hit of 1978, 'Everyone's A Winner', told us, we are all stars and we are all winners – no one needs to lose.

This mentality led to a belief that failure was a social construct by the Establishment. People believed they should feel good about themselves, and anyone that made them feel bad should be ignored or challenged. A colleague recently told me of a phrase graffitied on a wall in Paris during the 1968 student riots: 'I decree a permanent state of happiness.' Those well-meaning baby boomers shunned sadness and failure as unnecessary and avoidable – indeed, their revolution was not only against the Establishment, but against those who told them they could not do what they wanted, who told them they were not stars and not winners, or who told them they should not expect to feel good permanently.

Such an approach to life has had a deep and enduring impact on how we raise and educate our children. In education, there has been a trend to protect children, and even young adults in universities and colleges, from failure. Two researchers in the US – Stuart Rojstaczer, a former professor at Duke University, and Christopher Healy, a computer scientist at Furnam University – examined the inflation of grades in American colleges and universities. What they found was published in 2012.[3] Drawing on the grades of approximately 1.5 million students, their data showed that in 1960 (as in the 1940s and 1950s), C was the most common grade nationwide, while Ds and Fs accounted for more grades combined than did As. From the 1960s to the mid-1970s, grades across America rose rapidly, with Bs the most common and As the second most common grades awarded. By the mid-1980s, the awarding of As began to rise again, and by 2008, As were nearly three times more common than they were in 1960. As had become the most common grade across the board, awarded to 43 per cent of all students (see Figure 1).

The sentiment that 'everyone's a winner' settled firmly into our thinking from the early 1960s until today. We have stopped telling

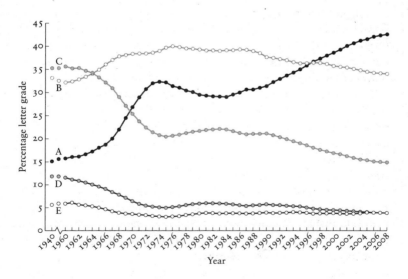

Figure 1. Distribution of grades at American colleges and universities as a function of time.

people when their work is not good enough, and we have stopped setting high standards, because students are failing to cope with criticism.

We have been heavily influenced by the self-esteem movement. According to this, to think positively about ourselves is the highest good and, if we do, we are likely to experience a range of positive consequences, ranging from better employment prospects, emotional stability and educational success on the one hand, to reduced chances of criminality and drug addiction on the other. It was the appeal of the self-esteem movement that led the Governor of California, George Deukmejian, in 1986, to fund a multi-year Task Force on Self-Esteem and Personal and Social Responsibility. Although the evidence for the perceived benefits of high self-esteem was either incomplete or unconvincing, in 1995 the National Association for Self-Esteem took over the mandate to promote a positive self-image among Americans, with the firm belief that helping citizens to think and feel good about themselves was central to a happy, prosperous society.

In a highly influential article published in 2003, Roy Baumeister and his colleagues evaluated the benefits of self-esteem and the virtues of the self-esteem movement.[4] They concluded that while people with high self-esteem claim to be more likeable and attractive, better at relationships and better at creating positive impressions on others, objective measures did not support those conclusions. Rather, people with very high self-esteem (commonly referred to as narcissists) tended to be charming at first, but in the long run alienated others.

People with high self-esteem are commonly believed to be better leaders, yet according to Baumeister and his team, this may be only indirectly related to self-esteem: people high in self-esteem show stronger in-group favouritism, which increases prejudice and discrimination but may also may make them more prototypical and therefore more often selected as leaders. While self-esteem may be strongly related to happiness, it does not prevent children from smoking, drinking, taking drugs or engaging in sex prematurely, as many at the time believed it would. Rather, they suggest that, if anything, self-esteem may increase the likelihood of experimenting with these things. On the whole, Baumeister and his colleagues reasoned that the benefits of high self-esteem probably fall into two categories: enhanced initiative and pleasant feelings.

In the end, promoting high self-esteem has fewer benefits than advertised. More concerning is the possibility that protecting the self-image of our children, rather than leading them to cope with failure and disappointment, may be doing more harm than good. As Jean M. Twenge, author of *The Narcissism Epidemic: Living in the Age of Entitlement,*[5] suggests, 'the "everybody gets a trophy" mentality basically says that you're going to get rewarded just for showing up. That won't build true self-esteem; instead, it builds this empty sense of "I'm just fantastic, not because I did anything but just because I'm here."'[6]

Considering that high self-esteem appears to have little benefit for greater overall well-being, and may even encourage narcissistic traits, you would think we might temper our efforts at promoting self-esteem. Yet, a study published in 2011 by Yalda Uhls and Patricia Greenfield suggests quite the opposite; we are promoting the value of self-importance to our children more than we used to.[7] The study selected the two most popular TV shows aimed at audiences of ten- to twelve-year-olds from 1967, 1977, 1987, 1997 and 2007. They

Value	1967		1977		1987		1997		2007	
	Means	Rank	Means	Rank	Means	Rank	Means	Rank	Means	Rank
fame	-0.44	15	-0.42	13	-0.59	15	-0.44	15	1.07	1
achievement	-0.08	10	-0.45	14	0.12	8.5	-0.08	10.5	0.97	2
popularity	0.39	5	0.46	5	0.25	7	0.39	6	0.87	3
image	0.64	3	0.64	2	0.32	4	0.64	3	0.83	4
financial success	-0.16	12	-0.48	15	-0.14	10	-0.16	12	0.77	5
self-centred	-0.05	9	-0.14	10	0.53	3	-0.05	9	0.73	6
power	-0.08	11	-0.17	11	0.38	5	-0.08	10.5	0.49	7
self-acceptance	0.45	6	0.58	3	1.06	1	0.59	5	0.04	8
physical fitness	-0.91	16	-0.36	12	-0.59	16	-0.91	16	-0.03	9
hedonism	-0.33	14	0.08	9	-0.30	13	-0.41	14	-0.06	10
community feeling	1.20	1	0.93	1	1.01	2	1.20	1	-0.10	11
benevolence	0.67	2	0.39	7	0.38	6	0.67	2	-0.26	12
conformity	0.06	8	0.43	6	-0.33	14	0.06	8	-0.79	13
security	0.31	7	0.27	8	0.12	8.5	0.31	7	-0.96	14
tradition	0.59	4	0.52	4	-0.20	11	0.58	4	-0.99	15
spiritualism	-0.33	13	-0.53	16	-0.22	12	-0.33	13	-1.03	16

Note: Data originally rated from 1 (*not at all important*) to 4 (*extremely important*). Means represent distance from grand mean for each decade.

Figure 2: Rankings of values by decade, ordered according to 2007 ranks

asked participants to rate how important certain values were in each of those shows. The different values are listed in Figure 2.

The value of community feeling (to be part of a group) was ranked the most important value communicated by the two most popular 'tweenage' television shows in 1967. Benevolence came second. By 2007 the desire for fame was the most important value communicated to pre-teens, while community feeling ranked 11th out of 16 and benevolence was down to 12th. This almost complete inversion of communicated values over a forty-year span is astounding. Children today are exposed to a completely different value system, one that is more focused on personal success and achievement, and less focused on our relations with others, than in the past.

Confirming this shift in values, a follow-up study by the same authors using pre-teenager focus groups asked participants to name the most important value for their future goals.[8] The ten- to twelve-year-olds in the study selected the statement 'The most important thing for your future is to be famous' as their number one goal, compared to other statements such as 'The most important thing will be to be really kind', or ' . . . to be part of a group'.

Consistent with the values being promoted through popular television shows, kids today believe the most important thing in life is to be liked by other people, and that this provides their main source of value.

Returning to the work of Jean M. Twenge and colleagues, in a review of research focusing on over 16,000 American college students between 1979 and 2006, they found narcissistic personality traits were on the rise.[9] Specifically, they observed a rise of 30 per cent in these traits in recent college students, compared to the average level of these traits observed in college students between 1979 and 1985. Although questions have been raised over the actual size of this increase and whether it is a trend that is specific to America, when considered together with the other studies above it does appear that people have become more focused on their own success, whether they are physically attractive, and whether others pay them sufficient attention.

We have become convinced our children need to feel good about themselves, to maintain a positive self-image. To this end, we seek to protect them from experiencing failure or disappointment in life, and instead we bolster their self-esteem. Yet, in so doing we are not making them better people or more proactive members of society; rather we are increasing their self-importance and self-entitlement.

THE RISE OF THE HELICOPTER PARENT

As parents, we seem to be hovering over our children more than before. A study conducted in 2004 by Liana Sayer and her colleagues confirms this suspicion.[10] She asked parents to fill in a time diary for a 24-hour period, indicating what their primary activities were over that time. Contrary to received wisdom, she found both mothers and fathers reported spending greater amounts of time caring for their children in the late 1990s than in the 'family-oriented' 1960s. What is especially interesting is that this increase in time spent parenting is occurring against a backdrop of increased workforce participation by women and an overall increase in the number of hours spent at work for both men and women. We are not spending any more time at home (and perhaps less), but when we are there we are putting in a lot of extra effort to be there for our kids.

However, this also means our children have less independence than we did when we were growing up. We don't let our children out of our sight anymore. How often do we let them walk to the local store on their own, or head off to the local playground unsupervised, or even just sit in the car for a moment while we pick up the dry cleaning? If we did, we would run the risk of being derided by those around us for failing to care properly for our children.

Regardless of your age, if you think back on your childhood, this change in parenting style (and what others expect of parents) is probably apparent. I grew up on a large property out in the bush. I recall spending a great deal of my time out of the house and 'down the back' among the old chook sheds (the property used to be a poultry farm), an old dilapidated house that was the original homestead on the property, and immense piles of what my mother called 'junk'. That 'junk' was the result of my father a) being in the demolition business, and b) having a tendency to hoard. The property was covered with bits of buildings that had been deconstructed but which my father planned to reconstruct one day; a number of old trucks and machines that had long since broken down; old cars that no longer worked; and any number of other fantastic piles of 'junk' that I spent my childhood foraging through. I would spend hours alone or with friends making cubby houses and otherwise occupying myself with very little parental supervision, among heavy machinery.

I also recall cycling to school when I was still in primary school. Along with some friends we would ride our bikes the entire 8 km. I remember being excited every day we had our bikes because I knew we would be riding home again. Of course, we would stop at a Milk Bar (the Australian version of the corner shop) along the way to get a drink and buy a mixed bag of sweets.

Those were good times, but they are very different from the experience of children today. Over the past forty years, there has been a significant decrease in the incidental exercise children are exposed to, such as walking to school.[11] More often than not we prefer to drive our children to school than let them run the risk of being unsupervised.

One reason we do this is because we think the world is simply not as safe today as it used to be. The thing is, there is plenty of evidence showing the world is actually safer than it has ever been. Take a recent book by Steven Pinker, the influential thinker and professor of psychology at Harvard University, titled *The Better Angels of Our Nature*.[12] Pinker has compiled an overwhelming amount of evidence to show the enormous decrease in the incidence of violence. This includes military conflicts, homicides, genocides, torture, criminal injustice, and the violent treatment of children, homosexuals, animals and racial and ethnic minorities. Pinker concludes we are probably living in the most peaceful time in human history. In the age-old fight between the angel and the devil on our shoulder, the angel is winning. According to Pinker, this is the result of our growing ability to detach ourselves from our immediate experience and reason from a more abstract and universal perspective than we have in the past. David Finkelhor, the director of the Crimes Against Children Research Center in New Hampshire, supports this. Perhaps the most reliable authority on sexual abuse and abduction statistics for children, Finkelhor has studied cases of abduction which are defined by three criteria: victims disappear overnight, or are taken more than 50 miles away, or are killed. These cases are a parent's worst nightmare and are also most likely to hit the national news. What he discovered is that such instances of abduction remain exceedingly rare and have not increased since the mid-1980s. In keeping with a general decline in crime, Finkelhor reports on his website that 'there have been some fascinating and encouraging developments in child victimization that have not received much publicity and run counter to popular perception. Certain types of abuse and crimes against children have been declining since the early 1990s.' Specifically, from 1990 to 2007, substantiated cases of child sexual abuse have declined 53 per cent, and physical abuse substantiations have declined 52 per cent. From 1993 to 2005, sexual assaults on teenagers decreased by 52 per cent. Other crimes against children aged twelve to seventeen years old have also declined, including simple assault (down 59%), robbery (down 62%) and larceny (down 54%).

The evidence is running counter to our intuition on these matters: instead of becoming more dangerous, the world is in fact becoming significantly safer. So why the helicopter parenting?

There are two possible answers to this question. The first is that while aggression, violence and crime are in decline, so is our trust in others. Robert Putnam, in his book *Bowling Alone*,[13] documents how people are less likely to rely on others than ever, marking a decline in 'social capital'. The notion of social capital reflects the idea that while I could use my economic capital (money) to buy a cup of sugar, I could use my social capital to ask my neighbour for a cup of sugar instead. Which raises the question – when was the last time you asked your neighbour for a cup of sugar? Putnam suggests this decline in social capital is linked to a reduced tendency to belong to social organizations (such as churches or rotary clubs). One of the implications of this decline in social capital is a decline of trust. If we don't trust that our kids will be safe in our communities, we don't leave them in the care of others.

Secondly, we are more aware of risk. Today, through our phones, tablets, computers and televisions, we are immediately informed of the kinds of events Finkelhor characterizes as newsworthy. When a child goes missing the whole country knows about it, and the more gruesome the circumstance the more often we hear about it. While crime is in decline, our exposure to it has vastly increased and the coverage of these matters is grossly sensationalized.

Children are no longer afforded the freedom and independence that we were because we are scared of the worst. We cannot stand to consider the possibility our children may be exposed to harm. By wrapping them in cotton wool, we are partly acting out our own fear of losing our children. When we think about something happening to our children it is our own pain that sends a shudder through our body – how could we go on if something like that happened? How could we ever forgive ourselves? This may be completely understandable, but our disproportionate fear of these painful outcomes has led us to create a developmental environment that is restricting our children in other ways, and causing secondary harm.

SOFTER AND SADDER

So what is the upshot of all of this? Why should we be so concerned our children are getting straight As and have never had to fend for themselves? While it hardly sounds like a problem, this environment could be making our children more depressed, more risk-averse and less resilient. By protecting them from failure, fear, pain and sadness, we undermine their ability to learn important life lessons, and ultimately to carve out happiness for themselves as adults. Take for instance a 2011 article in *The Atlantic* magazine titled 'How to Land Your Kid in Therapy'. Author, psychologist and parent Lorri Gottlieb recounts the various ways that protecting our kids may be doing them harm. Drawing on her experience of clients who had apparently perfect parents but were still suffering from depression, anxiety and other psychological ailments, she suggests there might be some unifying explanation. Could it be her patients were presenting with these issues in their lives not because they had been exposed to life's hardships but rather because they were entirely blinded to them? Referring to the work of Paul Bohn, a psychiatrist at UCLA, Gottlieb notes many parents seem to be doing everything they can to ensure that their children do not have to grapple with even mild discomfort, anxiety or disappointment. When those same children experience anything other than pleasant feelings or experiences as adults, they often think their whole world has fallen apart. They never learned to cope with these experiences as children and so have no idea how to respond to them appropriately (and independently) as adults.

It is instructive to consider our own responses to our children's pain. If you are like me, your first reaction when your child falls over and scrapes their knee is to run to them. When they get pushed or bullied by an older child in the playground, we want to intervene. However, intervening prevents children from learning how to manage their own discomfort or conflict. It also implies that the experience is unwarranted for their particular circumstances, and abnormal. Rather than viewing a scraped knee or playground conflict as incidental and largely a part of life, our response suggests it is terrible and out of the ordinary.

Dan Kindlon, a child psychologist, confirms this in his recent book *Too Much of a Good Thing*.[14] Drawing on his own data, he focuses on the link between over-protective parenting and the increased risk of depression. Kindlon argues that American children often lack the characteristics that are essential for well-being as a result of overprotective parenting. By protecting them from hardship and pain, parents have deprived them of the tools they need to mature, to become empathetic, to learn from failure and accept their flaws.

What is interesting is that we appear to understand and accept how this process of building resilience works in the biological domain. Mostly, we are all sold on the benefits of immunization. By injecting our children with a small amount of a pathogen we feel confident this will strengthen their immune system; it will build their biological immunity so their bodies will cope better with that pathogen in the future. And this is exactly how psychological resilience works. When we are exposed to manageable personal challenges, such as pain, failure or loss, we strengthen our capacity to cope with these experiences in the future. Mirroring the effect of pathogens on the physical immune system, our negative experiences in life literally build psychological immunity. We have to allow our children to experience challenging situations in order to build psychological resilience.

Ellen Sandseter, a developmental psychologist, has suggested that there may be evolutionary reasons for why children need to engage in risky play. In a paper published in the journal *Evolutionary Psychology*,[15] Sandseter posits that risky play not only allows children to experience exhilarating positive emotions, but also exposes them to natural fears. As the child develops and becomes more capable they are able to master these fears, which protect children when they are young but which limit them as they grow into adults. Contrary to what you might expect, children who were injured by falling from heights when they were between five and nine years old are less likely to be afraid of heights when they become adults. 'Paradoxically,' Sandseter suggests, 'our fear of children being harmed by mostly harmless injuries may result in more fearful children and increased levels of psychopathology.'

It was these concerns that led Wayne Bulpitt, the UK Chief

Commissioner of the Scout Association, to say that overprotective parenting and fears for safety are denying children a proper education as 'rounded individuals'. Reported in the *Daily Telegraph*, Bulpitt said, 'In Scouting our approach is that health and safety is very important but that young people have to be able to take risks in a supported and managed way in order to develop themselves. Where they get into difficulty in life is if they are protected in cotton wool.'

Together, the evidence indicates that when children are protected from experiencing difficult circumstances, it undermines their capacity to cope with those circumstances in the future. Critically, it also makes them less likely to seek them out because they feel less confident dealing with feelings of fear or uncertainty. Taking risks in life builds resilience and fosters an ability to face more risks in the future. Moreover, a failure to take risks is a hallmark of depression.[16] To be well-rounded, healthy and happy in life, we need to take a few risks.

Further evidence that being overprotected can make us less adventurous in life comes from data aggregated at the geographical level. Researchers sourced data from nearly 150,000 people, spanning all the major geographical regions of the world.[17] Specifically, they examined measures of risk-taking behaviour along with a number of indicators of hardship that people may face within each geographically distinct context. These included the homicide rates in each country, the Gross Domestic Product of each country, income inequality, infant mortality, life expectancy at birth and gender inequality. Of interest is that these indicators were all significantly related to each other: poor countries tend to have higher levels of income inequality and people living in these countries are exposed to more homicide, higher rates of infant mortality, lower life expectancy and greater gender inequality. For the purposes of this study, however, it meant the researchers could combine these various indicators into an overall measure of 'hardship'. They then used this measure (along with other indicators) to predict risk-taking behaviour. First, what they found replicated what we have known for a long time; as age increases, risk-taking behaviour decreases – older people are less likely to take risks. Yet, they also found that as hardship increases so does risk-taking behaviour. The more people had been exposed to difficult life circumstances, the more likely they were to take risks.

Now it is important to note that, as the authors of this paper argue, this may be because in uncertain environments where resources are diminished, taking risks is a more necessary life strategy. But, it is also true that those who were exposed to difficulties were more willing and able to take such risks. No doubt, taking many risks in highly uncertain environments is unlikely to be beneficial for well-being. Yet, taking few risks in highly certain environments is perhaps just as detrimental. Experiencing some hardship may not only push us to take risks, but it might prepare us for managing these risks. If you have faced worse, then risking a bad outcome is unlikely to feel as scary.

TRIGGER WARNINGS

This failure to develop psychological resilience has led to an emergent phenomenon within colleges and universities. Lately, it seems that students not only expect to get As or to avoid failure, they also expect to be protected from contradiction. In March 2016 students at Emory University woke to find messages all over campus written in chalk in support of Donald Trump. Rather than viewing this as something they disagreed with, as an indication that a diversity of opinions existed on campus, some students saw this as an act of intimidation. They responded with a protest – about forty or fifty students shouted in the quad: 'You are not listening! Come speak to us, we are in pain!'[18] They wanted university officials to recognize that being exposed to disagreeable opinions had caused them to suffer.

This inability to manage intellectual content perceived as challenging to one's own views, or as threatening to one's sense of personal comfort, has led to the spreading use of 'trigger warnings'. The idea is that it is important to warn people of specific content that could trigger past traumas, a practice that emerged particularly around issues of sexual assault. While this is not to be taken lightly, this protectionist philosophy has been growing into something altogether different and is now beginning to threaten our freedom of speech, and our ability to openly and honestly discuss different perspectives and ideas. Trigger warnings are now increasingly demanded and expected around college campuses and within lectures. This concern for being 'triggered' is even leading to professors needing to warn their pupils that *The Great Gatsby* portrays misogyny and physical abuse.

In response to this trend, Greg Lukianoff and Jonathan Haidt wrote a front-page article for *The Atlantic* which was titled 'The Coddling of the American Mind'. The authors make the claim that protecting college students from all potential distress may be linked to the increase in the rates of emotional distress reported by college students themselves. Lukianoff and Haidt draw on the most well-known form of psychological therapy – Cognitive Behavioural Therapy – in order to show that trigger warnings are in fact exactly the opposite of what we should be providing these teenagers with. Psychotherapy is built on the idea that gradual exposure, rather than avoidance, is the key to overcoming fears, anxieties and trauma. When an emphasis is put on the psychological potency of information, and people begin to expect they have a right *not* to be exposed to things that might upset them, a troubling pattern emerges. These individuals begin to live in fear of exposure, and, as was noted in Chapter 1, fear and in turn avoidance of things we do not like overstate their influence in our minds.

Of course, it is not only the fear that is so problematic; it is also the sense of entitlement to a life free of pain – a right not to be 'triggered' by life itself. This has led a swathe of professors at American universities to express their concern that the content of their courses is now being shaped, not by learning needs but by emotional needs. As Jeannie Suk Gersen, a Harvard law professor, wrote in the *New Yorker*, trying to teach rape law to budding lawyers in the current climate is like trying to teach surgery to a doctor who refuses to have anything to do with blood. As a prime example, she notes a case she heard of where one student asked their teacher not to use the word 'violate' – as in 'does this conduct violate the law?' – because the word was 'triggering'.

Our cotton wool culture has started to shape not only what is considered politically correct but what is ideologically tolerable. This has led people to refer to the young adults of the 2010s as the 'Snowflake Generation' – a reference to their inflated sense of uniqueness and their propensity for taking offence. This generation protests against their exposure to potentially triggering information, exhibiting their exceedingly narrow view of what the world should be rather than attempting to understand what it is. Of course, the downside is that this emotional mandate has a significant educational cost.

REMOVING THE PADDING

So what should we do about this? How can we remove the cotton wool and give the next generation the best shot?

There is now a groundswell occurring against the effects of mollycoddling. In a recent move, the University of Chicago now welcomes new students with a letter banning the use of trigger warnings or safe spaces where students can be protected from interrogation by others or expect not to be confronted by challenging and diverse opinions.[19] In the letter it states:

> Our commitment to academic freedom means that we do not support so-called 'trigger warnings,' we do not cancel invited speakers because their topics might prove controversial and we do not condone the creation of intellectual 'safe spaces' where individuals can retreat from ideas and perspectives that are at odds with their own.

This same trend is stretching all the way from colleges and risky ideas to primary schools and risky play grounds. In Swanson Primary School in New Zealand, there is a new rule for playtimes: there are no rules. Well, there is one rule, according to Principal Bruce McLachlan, which is 'don't kill anyone'. At playtimes, children as young as five can climb as high as they are able to, they are not required to wear shoes, and the teachers are not allowed to intervene unless they are explicitly asked to. When asked why he decided to create a 'school with no rules', McLachlan says that he got tired of saying no all the time – the idea was to start letting the children explore their own boundaries.

What has happened as a result of this 'no rules' policy? At the time this story was aired on the Australian TV show *Lateline*, only one incident involving a broken arm had been reported. Fearing a showdown with the parents, McLachlan was pleasantly surprised when they felt the injury was an important lesson for their son. The father said his son would now be more aware of his limits and that going beyond them can have consequences. Some might say McLachlan got lucky.

Beyond this one incident, however, the overall number of injuries had been reduced. Children were left to take responsibility for their own safety, and they did. Not only that, but once the children were back in the classroom, teachers reported they were more focused on learning. They were calm and engaged. Incidents of bullying had also been reduced since the new policy. Now that kids were learning how to do things for themselves, they were discovering how to get along with each other on their own terms.

Another, more surprising factor that emerged was the reduced use of iPods, iPhones and other gaming devices. The children were all outside engaging in play. This led me to reflect on an advertisement for the TV company Foxtel in Australia. It shows a mother telling her children to put down their devices and 'get outside'. She then recalls the various risks she took as a young girl and suddenly changes her mind: she tells them to sit down and watch a movie. It's a perfect pitch for Foxtel, but not for children's health and well-being.

A major factor that can either foster a willingness to let children take risks or fundamentally prohibit it is our increasing tendency to litigate. One of the reasons why the 'no cotton wool school' was possible in New Zealand is because of the free healthcare system. If children did hurt themselves, they could receive medical care without any associated costs, and this meant parents were unlikely to sue the school. In fact, it was exactly the threat of litigation that changed the landscape of playgrounds right across America. In 1985, the Nelson family in Chicago won a law suit against the Park District for compensation in the order of $9.5 million. As a toddler, their son, Frank, had fallen from the top of a slide, hit his head on the asphalt below and suffered brain damage, leaving him paralysed on one side and with speech and vision problems. It was a sad story indeed. The litigation arising from this case led to the creation of safe playgrounds that prohibited millions of children across America from facing risk, or overcoming challenges. Playgrounds were sanitized and stripped of their excitement due to a new motivating force – the wish to avoid litigation.

If we want to unwrap our children from their cotton wool, we need to challenge a few of our own assumptions about the world we live in. Experiencing pain is inevitable, and when we (or our children) do have those experiences we need to fight our inclination to see them as an injustice. Negative experiences provide a valuable pathway to develop resilience and when we protect children from pain, failure and loss, even from a diversity of opinions in life, it leaves them vulnerable and emotionally exposed.

PART TWO

Embracing Pain

Haruki Murakami owned a small jazz bar, which he sold in 1982 to devote himself to writing. It was a good decision. His first novel had won the Gunzo New Writers Prize in 1979 and some eighteen books later he won the prestigious Jerusalem Prize. In order to keep fit he began running, an activity that developed into an enjoyment of marathons and triathlons. His love of running also influenced his writing, resulting in a memoir covering his four-month preparation for the 2005 New York City Marathon. In his book *What I Talk About When I Talk About Running*, Murakami wrote the now famous quote: 'Pain is inevitable. Suffering is optional.'

Murakami understood the dangers of avoiding pain. He understood that by viewing pain as inevitable, by accepting it, we are then faced with the choice of suffering; that is, we can choose how we respond to pain. Yet, this quote does not entirely capture what pain meant to Murakami. It is true that he endured running in order that he might achieve fitness, but why did running then become such a passion? Why did he travel to faraway destinations just so he could endure the pain of long-distance running? Why did this activity, which was invariably arduous, become a driving source of inspiration and passion in his own writing?

What Murakami may not have appreciated is that pain was not an inevitable by-product of his running; it was the *very reason* why he ran.

3

Painful Pleasures

How strange would appear to be this thing that men call pleasure! And how curiously it is related to what is thought to be its opposite, pain! The two will never be found together in man, and yet if you seek the one and obtain it, you are almost bound always to get the other as well, just as though they were both attached to one and the same head . . . Whenever the one is found, the other follows up behind.

Plato, *Phaedo*

On 3 April 2015, eighty people converged on a semi-rural property north of Melbourne for an event called Hooked Up. The entry price for the sold-out picnic was $10 for observers. Twelve people, however, paid $200 so that they could participate in the event itself. For these twelve, participation meant having four de-barbed shark hooks inserted into their upper backs without painkillers by a qualified nurse. One-by-one they were then raised up to 5 metres off the ground by a pulley, dangling and swinging for up to 30 minutes.

This was the tenth Hooked Up event in seven years. For one participant it was his first time, and he was raised from hooks inserted into his upper back so that his feet were just off the ground for 5 minutes. Other examples from past events include a man who was raised face-up from six hooks inserted into his chest, solar plexus and hips. Another was suspended from his upper back with a 55-kilogram woman tied to a rope and hanging below him.

Participants for Hooked Up events come from all backgrounds and include school teachers, business people, police officers and a judge.

Their motivations vary as much as their backgrounds, ranging from sexual to spiritual to cathartic following personal trauma. Whatever their ultimate motivation, as the title of the news article reporting on this event in *The Age* – 'Extreme Body Piercers Hooked on Adrenalin'[1] – insinuates, participants in these events do it for the pleasure. As one reports, 'It hurt, but not as much as you'd expect. Once you're up, it goes a bit numbish and you sort of block it out. It's definitely fun, you feel the sunshine and the wind.'

Reading through this account it is easy to assume that participants in these events are mentally troubled, a little unbalanced, perhaps of questionable character. If you knew your child's kindergarten teacher enjoyed being hung up by hooks in their flesh would you feel comfortable leaving your children with them? It is for this reason that discretion and privacy are paramount at the Hooked Up events. People generally conclude that those who engage in this kind of pastime are deviants, not only because the activity is non-normative, but because it suggests something more sinister about a person's character – they are not the sort of people you trust.

This intuition that enjoying pain is not only abnormal but also immoral is strong. Yet, as with many of our intuitions, when you stop and really analyse this assumption, things become a little less clear. We get pleasure from many things in life, and seek out activities that provide this for us, so what is so different about getting pleasure from pain?

That date, 3 April 2015, was in fact Good Friday. I found it especially telling that even the organizer of the event, Bella van Nes, admitted it felt 'a bit naughty'. She said, 'If I still went to confession I'd have a lot to say, I think.' She claimed the overlap in dates was merely coincidental and more to do with convenience; yet she still found the overlap uncomfortable.

We generally refer to people who get pleasure in these ways as 'masochists', a term that carries derogatory connotations and reinforces the idea that such a form of pleasure is deviant and distasteful. Understanding the link between the experience of pleasure and the experience of pain, however, reveals that *all* of us get pleasure from pain. While we might see a sinister element in those who engage in more extreme or sexualized masochistic practices, there is a sliding scale including a great many commonly accepted forms of pain enjoyment. In these more normative versions we tend to forget that we derive our enjoyment precisely from the pain.

BENIGN MASOCHISM

Paul Rozin, the highly influential social psychologist, studies what he terms 'benign masochism', or the tendency to enjoy innately negative experiences. His interest in this phenomenon began with his own work on chilli consumption and the question of why well over 2 billion adults enjoy the 'burn' of chilli pepper in their mouths.

In order to further investigate these ideas, Rozin and his colleagues asked university students to answer questions about twenty-six different innately negative experiences.[2] The students indicated on a scale from o (not at all) to 100 (as much as I like anything) how much they liked each of the experiences. The experiences were very diverse, but could be grouped into eight different kinds. These were sadness (e.g. sad books or movies), burning sensations (e.g. spicy foods), feeling disgusted (e.g. squeezing pimples), fear (e.g. bungee jumping), painful temperatures (e.g. diving in a cold ocean), strong and innately negative tastes (e.g. whisky), feeling exhausted (e.g. running a marathon), and bitter tastes (e.g. strong coffee). Responses revealed that about 50 per cent of people on average reported enjoying most of these experiences at about the mid-point of the response scale. That is, a large number of people found these innately negative experiences at least moderately enjoyable.

Rozin and his colleagues then asked when people got their maximum enjoyment from these experiences. Remarkably, they found that between 25 and 60 per cent of respondents favoured the most extreme negative experience they could stand – that is, among other things, the most extreme roller-coaster ride, chilli pepper, massage or sad music they felt they could tolerate. This revealed that not only do a large number of people enjoy negative experiences, but their enjoyment peaks just before these experiences become unbearable.

The idea that people might get enjoyment from innately negative experiences has been referred to as *hedonic reversal* – literally a negative experience being reversed and becoming a positive one. One way to understand this is that people feel a sense of mastery or 'mind over body' when the brain realizes that the threat signal they experience does not reflect actual danger (such as the burning sensation in the mouth after chilli, or the fear felt during a roller-coaster ride). This feeling that we have overcome something that was potentially dangerous produces a sense of pleasure. Benign masochism characterizes the enjoyment of the conflict that arises when these simultaneous positive and negative emotions are activated. When we feel distanced from any actual threat, we are able to enjoy these otherwise threatening and negative experiences. This requires the capacity to cognitively override messages of threat being sent to the brain, such as when we tell ourselves that our sweaty palms and racing heart while watching a thriller are not a sign of real danger. There is little evidence that animals experience hedonic reversals because they do not have the same cognitive capacities as humans. For instance, Mexican dogs and pigs that regularly eat food with hot peppers (since they tend to be reliant on human leftovers) do not develop a preference for spicy foods, unlike the millions of Mexicans who do.[3]

Benign masochism explains the long queues at The Edge in the Eureka Tower in Melbourne. This popular tourist attraction involves walking into a glass box, which juts out from the edge of the tallest building in the city. The only thing separating you from the cityscape below is a couple of inches of glass. Every protective mechanism in your body is firing warning signals, telling your brain there is immense danger, yet you walk out into the glass box and stand there, finding a sense of enjoyment in the experience. You can do this because your cognitive abilities allow you to override the feeling of threat, telling your brain it is safe to take that first step onto the glass. This conflict between knowing you are safe yet feeling your body prepare for escape is what produces the thrill – an experience people are happy to part with $12 for.

A student of mine, Laura Ferris, investigated the various contexts in which people enjoy pain as part of her Ph.D. For one of her studies she attended Dark Mofo (Museum of Old and New Art: Festival of Music and Art), which is held during the winter month of June in Tasmania. As part of the festival there is a Winter Solstice Swim where more than seven hundred people venture into 11°C waters at sunrise ... naked. One of the questions we asked more than two hundred of the swimmers after they had participated in the event was to report how painful it was, and also how pleasurable they found it. Participants provided ratings on a scale from 1 (not at all) to 7 (very much so). This revealed that, unsurprisingly, the swimmers found the event painful, on average rating their pain as 4.15 out of 7. In parallel, however, they also rated the event as pleasurable, on average rating their pleasure as 6.13 out of 7. Not only did they find jumping into icy-cold water early in the morning to be highly pleasurable, that pleasure far outweighed the pain.

This tendency to indulge in the pleasure of icy-cold waters is not limited to naked art lovers. There is a tradition in the northern hemisphere – tracking all the way from America and throughout most of Central and Eastern Europe – of jumping into the ice-cold ocean on New Year's Day. This form of celebration may be an effective way to combat a hangover, but it is also looked forward to and repeated year after year. In Melbourne (as in many other places) there is the Icebergers swimming club, whose members get up at 5.30 on winter mornings and swim in the near-frozen ocean waters of the Brighton Baths, which often dip to 7°C. One of the oldest club members, aged seventy-six, reports, 'The reason I do it is, unfortunately, I don't know how to give up, it's addictive, I always feel really great after a swim.'[4]

Compare the participants in Rozin's research, along with those who get enjoyment from jumping into icy-cold ocean waters, to our Hooked Up masochists above. They are all masochists, only to varying degrees. While one person might like a lot of chilli on their food and enjoys skydiving or watching scary movies, another might enjoy only a moderate amount of chilli and prefers the sad moments in a romantic comedy to the gruesome violence of a horror film. Just as one person might enjoy being suspended by hooks in their skin and another person might not, from the perspective of benign masochism this difference now appears to be more a matter of degree than fact. We all regularly find pleasure in negative and painful experiences, yet it is when we see people taking this to extremes, and – crucially – beyond our own limits, that we feel the need to question their character.

HEDONIC REVERSALS AND
HEDONIC FLIPS

In 1997 Daniel Kahneman and his colleagues published a landmark paper titled 'Back to Bentham? Explorations of Experienced Utility' in the *Quarterly Journal of Economics*.[5] As with much of Kahneman's work, the paper aimed to convince economists that people's decisions cannot be predicted by simple objective calculations of utility. That is to say, very often people's decisions in life do not align with more objective measures of how much pleasure vs. pain they are likely to experience. Rather, argued Kahneman and his colleagues, there are two types of utility. One can be predicted by the known outcomes of a person's actions. For instance, my decision to go on holiday can be attributed to the known pleasure of relaxation and pampering – he calls this *decision utility*. The other type cannot be predicted by these known outcomes, but rather is determined by a person's subjective experience. For instance, we can imagine a case in which my enjoyment of a luxurious holiday might be very high because the other option, perhaps decided by the flip of a coin, was to do housework for two weeks – much less enticing. Relative to menial work, holidays are a welcomed pleasure. Yet, we can also imagine a case in which a luxurious holiday pales in comparison to the other option: wining $1 million. Relative to winning a fortune, a holiday feels particularly trivial. This capacity for our subjective experience to shape how much pleasure or pain we get from objectively similar events is what Kahneman calls *experienced utility*.

The point here is that context matters. Experiences are always relative, and rarely depend on the objective characteristics of the thing itself. Holidays are generally pleasant, yet they are *more* pleasant in a context where the alternative was more work, and *less* pleasant in a context where the alternative was receiving a small fortune.

In a study examining this effect of relative comparison on people's hedonic experiences, Barbara Mellers and her colleagues told volunteers that they would be playing a gambling game that involved a choice between pairs of gambles where they could win real money.[6] The volunteers were told that generally people come out ahead in these games, but in the event they lost more money than they won they would have to work off their 'debt' by doing menial tasks in the laboratory. In some trials participants earned $8 and in some trials they lost $8; however, these wins and losses were *framed* differently. Sometimes the alternative was to win $32 and other times it was to lose $32. Figure 3 below shows the study's results, where higher positive numbers indicate higher levels of disappointment. Across two different groups of volunteers, researchers found that when people won $8 they were more disappointed when the other possible outcome could have been to win $32 dollars. They were, however, less disappointed when the other possible outcome could have been to lose $32. In the same way, losing $8 was less disappointing when the alternative was to lose $32 than when it was to win $32.

As can be seen from the graph, *losing* $8 when the alternative was a possible $32 loss, feels about the same as *winning* $8 when the alternative was a possible $32 gain. What determined the volunteers'

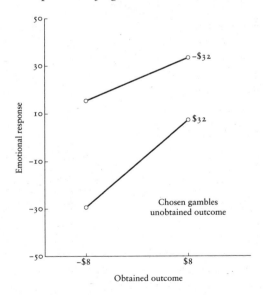

Figure 3. Graph showing the negativity and positivity of self-rated emotional response to winning or losing $8 when the alternative outcome was winning or losing $32.

level of disappointment was not only the absolute value of their win or loss (plus or minus $8), but rather the alternative against which their own outcome was contrasted (plus or minus $32). Losing can feel emotionally equivalent to winning, depending on the context.

Siri Leknes from the University of Oslo was also interested in how a given context could flip a negative experience into a positive one, but rather than showing this with losses and gains, she wanted to show that this could also happen with physical pain. The researchers recruited volunteers and exposed them to different levels of pain by applying heat to the inside of their forearms.[7] Consistent with the findings of Barbara Mellers, they found when a moderate level of pain was contrasted with a *less* painful experience (a warm sensation), the volunteers rated it as −1.2 on a scale from −5 (very painful) to +5 (very pleasant). However, when the same moderate level of pain was contrasted with a *more* painful experience, the volunteers gave it a +1.1 on the same scale, indicating their experience of the moderate stimulus had crossed over to become a mildly pleasant feeling. This finding illustrates what is referred to as a *hedonic flip*. The effect of a particular relative comparison can literally transform pain into pleasure.

This same hedonic flip can be observed in this account of an eczema sufferer, Dr John Launer, reporting his experience, which he titled 'The Itch', in an international medical journal:

> I have just run the hot water tap and put my hands underneath it, with the water as hot as I could bear for as long as I could bear. The water was probably hotter than most people could stand, certainly beyond the temperature to cause pain. That was why I did it. I have been trying to reach the pain threshold in order to 'crack' the itch from my eczema.[8]

Consistent with the experience of the Norwegian volunteers above, Dr Launer's account reveals the burn of hot water can feel rewarding when relieving an even worse experience. Of course, rather than using hot water many people will scratch their skin until is it is sore and bleeding. These kinds of relative contrast effects, which turn pain into pleasure, are far more frequent than we might at first recognize.

So, there are two ways that pain can be turned into pleasure. The first is what Paul Rozin called *hedonic reversals*, where people get enjoyment from potentially threatening experiences within a safe context – a kind of 'mind over body' mastery. We receive the benefit of 'thrill' rather than a 'threat'. The second is what Leknes refers to as a *hedonic flip*, where a negative experience may become pleasurable when compared to a more intense negative experience. From this we can see that the hedonic value of an experience is largely relative – it depends on what the alternative is. This finding is perhaps captured by the sentiment that we are often glad to receive the lesser of two 'evils'. Yet, this effect of relative value is strong enough such that, at times, it can lead us to experience a loss as similarly pleasurable to a win that is the lesser of two 'goods'.

The concept of benign masochism demonstrates that many more of us get pleasure from pain than we might realize, and as a recognized form of pleasure-seeking it is far from relegated to bizarre subcultural practices; rather, it is a part of our daily lives. Hedonic reversals and hedonic flips show us *how* our experience of pain can be shifted through relative contrasts and therefore *why* innately unpleasant experiences may sometimes feel pleasant and enjoyable, further blurring the distinction between pleasure and pain. Yet, there is another even more important link between these two experiences. To understand this we need to turn our attention from how a painful event itself can become pleasant, to what happens after pain.

THE PLEASURE OF RELIEF

In a widely cited 1974 paper, two psychologists, Richard Solomon and John Corbit, describe the phenomenon of *opponent processes*.[9] They begin by providing a fictitious story of a woman who discovers a lump in her breast.

> A woman at work discovers a lump in her breast and immediately is terrified. She sits still, intermittently weeping, or pacing the floor. After a few hours she regains her composure, stops crying and begins

to work. At this point, she is still tense and disturbed, but no longer terrified and distracted. She manifests symptoms usually associated with intense anxiety. While in this state she calls her doctor for an appointment. A few hours later she is in his office, still tense, still frightened: She is obviously a very unhappy woman. The doctor makes his examination. He then informs her that there is no possibility of cancer, that there is nothing to worry about, and that her problem is just a clogged sebaceous gland requiring no medical attention. A few minutes later, the woman leaves the doctor's office smiling, greeting strangers, and walking with an unusually buoyant stride. Her euphoric mood permeates all her activities as she resumes her normal duties. She exudes joy, which is not in character for her. A few hours later, however, she is working in her normal, perfunctory way. Her emotional expression is back to normal. She once more has the personality immediately recognizable by all her friends. Gone is the euphoria and there is no hint of the earlier terrifying experience that day.

This kind of experience, where a negative event gives way to a sense of relief that itself turns into something pleasant, is evident in many everyday contexts. Take, for instance, eating or drinking when you are very hungry or very thirsty. The reward value of food and water in these cases is substantially increased by the sense of relieving a pent-up hunger or thirst. Also, recall the example of the enjoyment gained from a warm beer and smelly hot spring in Chapter 1 – it was the discomfort and effort of hiking a long distance prior to this experience that made it so enjoyable. Opponent process theory describes this effect and, as such, how a feeling of discomfort can increase our experience of subsequent pleasure – more so than if we had never had any discomfort in the first place. This can be seen in Figure 4 opposite, where State A is the negative event (cancer scare) and State B is the positive response (euphoria). At the bottom of the figure we can see the thick black line indicating when the negative event begins and ends.

To illustrate, imagine for a moment jumping into a cold shower. Focusing first on the left panel of our figure, at the onset you suddenly feel a shock of cold water (this is peak A), and then over time you adapt to the cold and the initial shock of the cold water subsides.

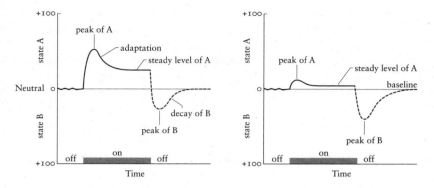

Figure 4. The graph on the left depicts the timeline of how opponent processes work. State A is the initial state which peaks at the beginning and then the individual becomes accustomed to it. When State A ceases the individual moves into State B (the opponent process) and this peaks before returning to the baseline. The graph on the right shows how this process changes once it has been repeated on multiple occasions. State A is less intense but State B (the opponent process) becomes more intense.

Next, you turn the shower off and the cold shock stops. Now instead of experiencing State A (cold shock) you are experiencing its opposite, State B (pleasant relief). Like the initial shock of cold this peaks quickly (peak B) and then subsides over time, eventually returning back to the baseline.

Opponent process theory shows how a negative experience can produce a positive response. It feels good when the pain stops, more so than if we had never had it in the first place. This may increase the value of rewards such as food or water, but what is crucial here is that the experience of relief is a reward in itself.

A study conducted by Joseph Franklin at the University of North Carolina examined the emotional state of relief.[10] He and his colleagues recruited forty volunteers and attached electrodes to their left bicep. Across several different trials the researchers administered an electric shock to the volunteers and then presented them with a sudden loud noise through a set of headphones. This allowed the researchers to examine how the experience of pain cessation (relief) might affect emotional reactions to the loud noise. In order to examine the volunteers' emotional reactions they measured small changes in muscle activity around the eyes and ears. When we experience a sudden negative reaction, for example to a loud noise, we tend to blink and this can be used as a measure of negative affect. The researchers measured how reactive this reflex was when the volunteers heard the sudden loud noise. They also measured the activity of the postauricular muscle, which is located behind the ear and is activated by positive emotional experiences. Together, this allowed the researchers to measure how much negative emotion (eye blink) and positive emotion (postauricular activity) participants experienced after being shocked and in response to the loud noise. They found that compared to just being presented with the loud noise on its own, when participants had just been shocked and were in a state of relief, the reaction of their facial muscles to the sudden loud sound indicated they were experiencing more positive emotion and less negative emotion – they were less likely to blink and had greater postauricular activity. Furthermore, the researchers also found that this effect became stronger when the intensity of the electric shocks was increased.

Mostly we think of relief as just the absence of a negative feeling – the pain is gone and my relief is simply the absence of that pain. Yet, this is not what relief is comprised of. Relief is itself a positive feeling, in addition to the absence of negative feeling. Relief is a rewarding experience, but we of course can only experience relief if we first endure some discomfort.

The question, however, is whether it is worth enduring cold, hunger, thirst or pain to achieve this short-lived pleasure surge? There are two reasons why it is. The most important reason will be made clear in the last section of this chapter – endless pleasure is really just a hedonic mirage – but for now we can answer this question by understanding the way opponent processes work. A well-known fact is that when we experience something repeatedly we tend to adapt to it. This is true of both positive and negative experiences. Yet, it is not true of the opponent process that results; in fact, this becomes stronger. So, while over time we adapt to the unpleasantness of doing something we dislike, we tend to experience more pleasure as a result. This is evident when comparing the graph in the right-hand panel in Figure 4 to the graph in the left-hand panel. State A (the negative event) becomes less prominent over repeated instances, yet the opponent process State B (the positive outcome) becomes stronger.

The first time you go for an early-morning jog (fitness and other factors aside) you will experience a great shock, but over time you get used to what it feels like to be out pounding the pavement as the sun rises. When you return home, however, you will retain that same euphoric feeling, and this may even increase the more you repeat this activity each morning.

Of course, the same is true for pleasant experiences; we adapt to them over repeated instances, but the negative opponent process (the displeasure that follows) becomes stronger over time. This perspective certainly suggests that developing a routine which involves repeated unpleasant experiences is more likely to increase our overall levels of pleasure than is a routine which involves repeated instances of pleasure.

Neuroscientists studying the brain's reward systems have provided some explanation of how these opponent processes operate. Painful and pleasurable experiences share similar neural mechanisms. Researchers working with both rats and humans have found that at times painful experiences can in fact trigger the brain's reward circuitry – that is, pain can trigger the same chemical responses that are involved in our ability to experience pleasure. Two key neurotransmitters that are triggered by pleasure and pain are *opioids* and *dopamine*. Both are released in the brain when we enjoy a pleasurable experience. For instance, enjoying a nice meal can cause activation of both opioids and dopamine, but – importantly – they play different roles: opioids are for 'liking' and dopamine is for 'wanting'. That is, opioids facilitate our enjoyment of something and dopamine serves to increase our desire for more. In fact, it is exactly this combination of neurotransmitter activation that is involved in drug addiction (and for that matter, almost any type of addiction). Drugs serve as a powerful trigger for these neurochemical processes, which in turn feed into our liking for the drug and our desire for more.

These shared neural mechanisms provide important insight into how a feeling of discomfort can give way to an experience of pleasure. Consider the example of the elderly Iceberger from Melbourne who reported not only feeling addicted to swimming in the near-frozen waters of the Brighton Baths, but also reported feeling really great after her swim. Much like a drug, the cold water activates the release of dopamine, which compels her to return for a freezing dip day after day. In addition to activating the release of dopamine (the 'wanting' neurotransmitter), the cold waters also activate the release of opioids (the 'liking' neurotransmitter). As I will detail below, there is good reason for this – both opioids and dopamine are especially effective in reducing and responding to pain. Yet, the swimmer's brain continues to release opioids after she emerges from the water, and as the sting of the icy-cold water begins to subside, this overshoot of opioid release now gives way to an experience of pleasure. It is exactly this pattern of opioid release in the brain that is thought to underpin the opponent process that I described above: when an unpleasant experience ends, it gives way to an increase in pleasure.

One domain in which this pattern of pain giving way to pleasure has been noted is in the case of the 'runner's high'. This is the phenomenon where runners often report an experience of euphoria after an intense episode of running. In a 2008 study, Henning Boecker and his colleagues examined the underlying neurochemistry associated with this phenomenon. They recruited ten trained male athletes from a local running and sports club in Germany and had them complete a two-hour run. Both before and after the run they were able to measure the levels of opioids released into the brain. As expected, they found the athletes had higher levels of opioids in their brains after the run compared to beforehand. This demonstrated that running had led to an increased release of opioids into the brain. Most importantly, however, they also found that this increase in the levels of opioids released was positively related to the athletes' self-reported feelings of euphoria after they had finished the run. This was the first evidence for the long-held idea that opioid release was responsible for the so-called 'runner's high'. Engaging in strenuous activities can serve to trigger the release of opioids, but when the activity stops these neurochemicals remain in the brain, producing the opponent process of euphoria.

Together the evidence shows how painful, stressful and uncomfortable experiences can build pleasure in our lives. Such experiences give way to the sense of relief and underlying this is a neurochemical process, which (as I describe below) helps us to cope with our discomforts, but, when they stop, gives rise to an experience of pleasure.

PAIN-INDUCED ANALGESIA

Stressful, effortful or unpleasant qualities of a stimulus activate the release of dopamine and opioids in the brain, which in turn can lead to the experience of pleasure or even euphoria when the negative event ceases, but this is not the primary reason for the neurochemical response. Rather, the brain responds in this way because both opioids and dopamine help us to cope with, and respond to, pain. On the one hand, opioids function as a natural analgesic; they are like the brain's own fast-acting paracetamol. This is why people often do not feel the discomfort of an injury straight away. When we graze a knee or break a leg, our brain is flooded with these neurochemicals and they camouflage our experience of pain. It is often only afterwards that we start to realize how much it hurts. This same response can be observed in other negative events, such as social stress. On the other hand, although dopamine may also have an analgesic function, more recent research suggests it responds to the threat of discomfort or unpleasantness by facilitating effective decision-making. Specifically, it helps us to decide whether to continue to endure the unpleasant event (perhaps in order to get a reward) or to withdraw and escape (perhaps in order to prevent greater loss, such as an injury).[11]

Researchers have found that when people are not able to release enough of these neurochemicals they are less able to cope with painful experiences. They have also found these same individuals experience more sadness and depression; i.e., when the brain cannot release enough opioids, people tend to feel more sad and depressed. This is not dissimilar to the role of the neurochemical serotonin in depression. Drugs that aim to alleviate depression are called Selective

Serotonin Re-uptake Inhibitors (SSRIs), which enable higher levels of serotonin to remain in the brain, reducing feelings of sadness and depression. A failure to release a sufficient level of opioids in the brain has a similarly depressing effect on our moods.

This finding further supports the idea that distinct types of pain – physical and emotional – are regulated by similar neural systems. Such a possibility is underpinned by a large body of work showing an overlap between the experience of 'social' pain – when we are rejected or excluded by others or experience loss – and 'physical' pain.[12] This research has shown that not only are these experiences both referred to with common adjectives – such as people describing their experiences of rejection as feeling 'hurt' – but both also activate similar areas of the brain and produce similar cognitive, emotional and behavioural responses. In fact, the brain regions associated with pain are also involved in the detection of threat more generally, and are activated in situations that evoke feelings of fear, anxiety and negative emotion in general.[13]

Consistent with this argument, researchers have found that acetaminophen (paracetamol), commonly used for physical pain such as injury, muscle soreness or headaches, can also reduce 'hurt feelings' associated with social rejection or ostracism.[14] In this study, researchers recruited volunteers and gave them one 500 mg pill immediately after waking up each day, and another before going to sleep, for three weeks. The volunteers were randomly assigned to either the experimental condition where they were given the acetaminophen or to a control condition where they were given a placebo. Each evening the volunteers were asked to indicate on a 5-point scale whether the following questions were 'not at all characteristic of them' (1) or whether they were 'extremely characteristic of them' (5):

1. Today, my feelings were easily hurt
2. Today I was a sensitive person
3. Today I was 'thick skinned'
4. Today I took criticism well
5. Today, being teased hurt my feelings
6. Today, I rarely felt hurt by what other people said or did

For the volunteers who were given acetaminophen, their hurt

feelings declined significantly over the three-week period of the experiment. This was compared to the volunteers who were given the placebo, whose hurt feelings did not decrease at all. Aiming to provide more evidence for this finding, researchers ran a second experiment where they gave acetaminophen or placebo pills to volunteers, again for a three-week period. These volunteers were then invited to come and participate in an fMRI brain-scanning session which would allow observation of the effects of social pain on the brain. The volunteers were introduced to a ball-throwing game and played this while in the fMRI scanner. They saw the image on a computer screen and were told they could pass the ball to two other volunteers, who would then either pass it back to them or pass it between each other. The 'other volunteers' were in fact not real, and the game was pre-programmed to deliver two different kinds of social experience. In the first round of the game the participants in the study received the ball as often as the other two 'volunteers' did – this was the control condition. In the second round they received the ball three times and then did not receive it again, believing that the other participants were in fact excluding them – this was the social pain condition.

Results supported those of the first experiment. Volunteers who had been given the acetaminophen showed less activation in the regions of the brain involved in registering pain compared to those who were given the placebo. These two experiments show that pain arising from social rejection can not only be reduced using analgesics, but also that this reduction in 'hurt feelings' is related to reduced activation of the areas of the brain involved in processing physical pain.

Although there is emerging evidence to suggest this neural overlap may be less specific to these two types of pain than previously thought, the notion that these experiences share many characteristics, including similar patterns of neural activation, with a broad range of negative experiences – pain more generally defined – remains uncontested.[15] Such overlap across different types of pain also provides insight into how one kind of pain may help to reduce another. The experience of physical pain can serve to reduce our experience of negative emotion via activation of shared neural circuitry – specifically, the release of opioids. This provides useful insights into why people often resort to the use of physical injury, such as in the case of self-harm, to help regulate their negative emotional experiences. The pain in these cases may make them feel better because it activates the release of helpful neurochemicals. Importantly, however, pain does not *need* to occur in the context of harm, and more frequently we seek to regulate our negative emotions through heading to the gym or going for a run. These experiences are also likely to be more effective in regulating our emotions, as they do not involve physical harm but still can produce the same positive effects through endurance and effort.

Evidence for this link comes from a study by researchers at the University of Rochester, focusing on the effects of running and weightlifting on clinical depression.[16] The study focused on forty female volunteers diagnosed with depression. They were split into different groups and completed different tasks over an eight-week period. The volunteers in the 'Track' group walked or ran round an indoor running track four times a week. The 'Universal' group used a Universal Exercise Machine and completed a standard weightlifting programme four times a week. Finally, a 'Waitlist' control group were simply told their participation in the study had been delayed and they did nothing (this is a standard approach to creating a baseline). As you can see from the graph below, both the running and the weightlifting led to significant reductions in depression, compared to the waitlist control group.

The findings show us that enduring physical discomfort can help to regulate our negative emotional states, yet this is not a one-way process. Rather, if physical pain can reduce emotional turmoil, then it is possible this relationship could operate the other way too: negative social encounters or emotional states might also lead to a reduction in physical discomfort.

This possibility was examined by a group of researchers at the University of Toronto.[17] To test the theory they brought volunteers into their laboratory and asked them to interact with another person. The volunteers thought this other person was another participant, but she was in fact part of the research team – a trained actress who was instructed to either be 'cool, standoffish, and uninterested' or 'warm, friendly, and validating'. The researchers then induced pain

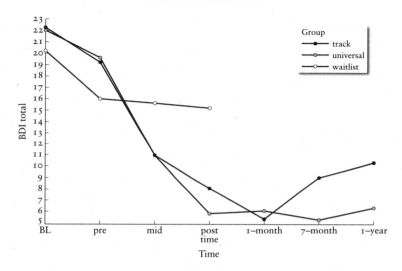

Figure 5. This shows scores on the Beck Depression Inventory (BDI) – a common measure of depressive symptoms – for participants in the different groups over the period of the intervention. By mid-way though the intervention differences were already emerging, with people track running or weightlifting experiencing fewer depressive symptoms compared to the 'waitlist' control group. This reduction in symptoms remained for up to a year post-intervention.

in the volunteers using a machine applied pressure to one of their fingers. What they found was the volunteers who had received the 'cool, standoffish, and uninterested' treatment reported a higher tolerance to physical pain compared to those who received the 'warm, friendly, and validating' treatment. Critically, volunteers in the 'warm' group reported a similar level of pain to others who did not have any social interaction at all, suggesting it was especially the negative interaction that reduced the experience of pain. It is important to note here that these effects were observed with just a mildly negative social encounter, demonstrating that even such minimal variations in our social world are sufficient to impact on our experience of physical pain.

The concept of pain-induced analgesia reveals the real reason why such events trigger the release of opioids and dopamine; the function of these neurotransmitters is to help us respond to, and cope with, pain. Furthermore, these analgesic effects can also spread from one type of unpleasant experience to other unpleasant experiences, and one reason for this is because our myriad negative experiences (be they physical, emotional or social) are regulated by similar neural responses in the brain.

As we saw above, the pattern of neurotransmitter activation responsible for pain-induced analgesia is also the reason why stress and pain can increase subsequent pleasure: opioids released in response to these negative experiences increase our subsequent feelings of enjoyment and positivity. Again, the reason is that both pain and pleasure are regulated by this same neural response. Of course, this also means pleasure should be effective in reducing pain, and indeed we know that it is. Experiencing the joy of social support is very effective in reducing the unpleasantness of physical injury (as we shall see in Chapter 5), and seeking out more positive and pleasurable events in life is a common and effective treatment for depression. In this way, pleasure also provides an important avenue through which we can enhance our enjoyment in life and better manage our negative experiences. Yet, pleasure alone cannot achieve these positive outcomes.

HOW PLEASURE TURNS TO PAIN

Opponent process theory not only helps us to understand how pain may turn into pleasure; it can also help us to understand how pleasure can turn into pain. The same rule applies to initially pleasant experiences (State A, see Figure 4 above) which then serve to produce an opponent process – a less pleasurable experience (State B) but which eventually returns to our normal baseline. Richard Solomon and John Corbit provide an example of this kind of experience.

> A couple have begun sexual foreplay, and it is quite pleasurable. After a few moments of a constant level of mutual stimulation, the pleasure decreases somewhat. Normally this decline would elicit behaviour calculated to increase the intensity of mutual stimulation and to maintain

the high level of pleasure. Unfortunately, at that moment a telephone rings. One partner leaves and goes into the other room to answer it, and the other partner lies alone in bed. The abandoned partner experiences a quick decline of pleasure, then becomes tense and irritated, and strongly craves a resumption of the sexual stimulation. Time goes by, however, and the other partner does not return. Finally, the abandoned partner gets out of bed, absentmindedly turns on the television set, and becomes absorbed in a news broadcast. Gone is the irritability and intense craving. There is no hint, in overt behaviour, of the pleasurable sexual experience a few minutes ago. A type of dispassionate normality now pervades.[18]

As you can see from this example, the initial State A is pleasurable and when it stops there is a corresponding State B that is not pleasurable. Again, this second state peaks, but then returns to normal. This example provided by Solomon and Corbit is consistent with research showing post-sex sadness is real. In a study that looked at women and men across four different countries (Brazil, Canada, Norway and America), it was found that approximately 40 per cent of participants experienced post-coital blues 'sometimes', and about 20 per cent experienced it 'often'.[19]

In another paper describing these opponent process effects, subtitled 'The costs of pleasure and the benefits of pain',[20] Richard Solomon describes an informal experiment he conducted where he went into the baby nursery of the Philadelphia General Hospital and presented a nursing bottle to several sleeping babies who were about twelve hours old. He notes such babies are not usually hungry or thirsty as they are still digesting a large quantity of amniotic fluid. Nonetheless, Solomon wiggled a nursing nipple into their mouths with a sweet milk solution, which woke them up and they began to suck. After allowing them to suck for a minute he withdrew the bottle. Not surprisingly, after about 5–10 seconds the babies cried for several minutes and then went back to sleep. Again, the same process is evident here: when a pleasant event suddenly ceases, the babies experienced a negative emotional state, albeit one that resolved after a period.

This opponent process mirrors its opposite: just as the cessation of a negative experience can induce the experience of pleasure, the cessation of a positive experience can reduce it. What is less clear in this case is how the underlying neurochemistry might contribute to this effect. While 'opioid overshoot' can lead to feelings of euphoria when pain ceases, it is less clear how opioid or dopamine release in the case of pleasure might activate subsequent negativity. Nonetheless, we can be sure that there are boundless examples of pleasant experiences increasing feelings of displeasure. Just take shopping as an example. Whether it is nice jacket or a shiny bicycle, new purchases feel great. Bringing the new possession home makes us feel happy and excited. Yet, it does not take very long for this feeling to wear off and by that evening we may be more worried about the impact on our budget than we are excited by the new purchase itself.

Such an effect was documented in a study examining the emotional responses of consumers.[21] Researchers asked consumers about the emotions they experienced when contemplating an important purchase, while shopping for an important purchase, and when using the product after it was purchased. What they found was that some consumers (those who value materialism) experienced more positive emotions prior to a purchase, but a decline in positive emotion after the purchase had occurred. This parallels the crying babies or the abandoned sexual partner above.

Drawing on the logic of opponent processes, we would also expect that as with initially negative events (recall the example of jogging in the morning above), over repeated instances we begin to adapt to the initial positive experience (e.g. the joy of shopping) but the negative opponent process becomes stronger (e.g. the sense of regret). This certainly suggests that the concept of 'retail therapy' has its limits, and that seeking endless pleasure not only does not work, but works less and less over repeated instances.

Although the above examples rely on an underlying emotional opponent process to understand how a positive event may lead to a subsequent negative experience, there are also other ways that seeking pleasant experiences might undermine our enjoyment in life. Gus Cooney from Harvard University examined an underlying social explanation for why extraordinary experiences might have unforeseen costs.[22] His research team started with the observation that people often seek out extraordinary experiences in life. As an example, six hundred people have now paid a minimum of $250,000 for a seat on the world's first commercial spacecraft, soon to be launched by Virgin Galactic. This experience will be truly extraordinary, and will remain so until such journeys become commonplace in the not-so-distant future. There are several reasons why people might be prepared to part with so much money for this experience. One is that it will clearly be an amazing experience. Another reason is that, as with much of what we do in life, we enjoy relating these experiences to others. Just as a new car is as much a status symbol as it is enjoyable to drive, having rare experiences can make us seem like interesting people others want to know. From this perspective, having exclusive experiences may also have social benefits. Cooney and his colleagues wondered, however, whether these six hundred people may be sorely mistaken.

To test this idea, Cooney, along with psychologist Daniel Gilbert, recruited volunteers to come into their laboratory at Harvard in groups of four. The volunteers were told they would be watching one of two videos. One video was referred to as the '4-star video' and the other was referred to as the '2-star video'. The volunteers were told these star ratings had been determined by the feedback of other students (which was true). They were then each assigned a video to watch, with three volunteers assigned to the 2-star video, the 'ordinary experience', and one volunteer assigned to the 4-star video, the 'extraordinary experience'. Everyone was made aware of what the other group members were going to watch before they went into separate cubicles to watch the videos. After the 10-minute videos had finished they were then escorted to a larger room where they sat around a table and were told to talk among themselves for 5 minutes. After that they were escorted back to their separate cubicles.

During this procedure the researchers asked the participants to provide a number of ratings. At the beginning of the study, and before watching the video, they rated how they were feeling on a scale from 0 (not very good) to 100 (very good). They also provided this same rating at the end of the group interaction, in addition to another question, which asked them how included they felt during the interaction on a scale from 0 (included) to 100 (excluded). Ratings revealed that at the beginning of the experiment, those who watched the 4-star video – the extraordinary experiencers – rated their feelings as 68.71 out of 100. This was very similar to those who watched the 2-star video – the ordinary experiencers – who rated their feelings as 68.22 out of 100. After watching the video and then having the 5-minute interaction, the extraordinary experiencers rated their feelings as 53.26 out of 100, while the ordinary experiences rated their feelings as 64.37. This demonstrated that having the extraordinary experience led to a decrease in happiness, rather than an increase in happiness. The reason for this was that the extraordinary experiencers also felt more excluded, rating their level of exclusion as 80.47 out of 100, compared to the ordinary experiencers, who felt relatively more included, rating their exclusion as 51.00 out of 100. Using statistical analyses, the researchers could demonstrate that ratings of exclusion explained the change in happiness, meaning the extraordinary experiencers felt less included and therefore less happy.

This study provides timely advice for the six hundred people still waiting for their extraordinary Virgin Galactic flight. Sell your ticket! It will probably *not* make you the star of dinner parties, and will more likely make you feel socially isolated and dissatisfied.

Whether it is an emotional opponent process, or the unforeseen social costs of extraordinarily pleasant experiences, the evidence suggests that seeking pleasure in life may have several downsides, leading to dissatisfaction. Yet, beyond inducing subsequent feelings of displeasure, could repeated indulgence in innately pleasant events change the value of that event itself?

An especially clear example of just such a process is when we indulge in something pleasant, but beyond the point at which we feel satisfied. Take, for instance, chocolate. One of the most immediately rewarding foods – it is high in both fats and sugars – it can quickly become a source of pain if we overindulge. Dana Small, currently at the Yale School of Medicine, and colleagues examined the effects of eating chocolate beyond satiety.[23] Specifically, they asked nine healthy volunteers (who reported they were chocoholics) to eat a piece of chocolate by letting it melt in their mouth. They then asked them to answer two questions using the rating scale below:

1. How pleasant or unpleasant was the piece of chocolate you just ate?
2. How much would you like or not like another piece of chocolate?

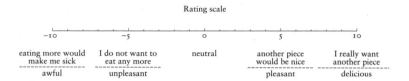

Rating scale

eating more would make me sick	I do not want to eat any more	neutral	another piece would be nice	I really want another piece
awful	unpleasant		pleasant	delicious

It is perhaps not all that surprising to see that, after each piece of chocolate, the pleasure people could derive steadily reduced. In fact, after the fourth piece of chocolate their ratings turned from indicating the experience was pleasurable and they really wanted another piece, to indicating the experience was awful and eating more would make them sick. This is an experience many of us have had after consuming one too many Easter eggs.

Of course, you might simply recommend that these volunteers should have stopped eating chocolate before it became painful – perhaps at the point when they started to feel satisfied. This would certainly avoid the feelings of disgust they experienced, but it would not provide much direction on what they should do next. Should they turn to something else enjoyable, such as watching television, or should they perhaps try hard to concentrate on their studies first? Watching television would be much more satisfying after expending effort on studying, rather than after their fourth piece of chocolate.

I would suggest that this same effect is true of all types of pleasure, whether it be time spent with friends, enjoying a natural landscape, or receiving a relaxing massage. These things can make us happy, but only if they are contrasted with something else. As I mentioned previously, Aldous Huxley envisioned a *Brave New World* where people could eradicate their pain and experience just pleasure alone. His story was not one of paradise found, however; it was about a dystopian future he wanted to warn us all against. A life full of pleasure quickly turns to disappointment – a banal mundanity, a numbness that leaves us feeling flat, depressed and sick to the pit of our stomachs.

4

Getting Tough

[To confront pain] is a struggle to be independent, tough, in charge of one's own life.

Arthur Kleinman, *The Illness Narratives*

On 16 April 2007, Seung-Hui Cho left his dorm room in Harper Hall at Virginia Polytechnic Institute early. In possession of a firearm, he made his way into West Ambler Johnston Hall student residence soon after 6.30 a.m., entered the dorm room of two female students, and shot them on the spot. After returning to his own room, deleting emails and removing his computer hard drive, he then went to a nearby post office and mailed his personal writings and video recordings to NBC News. A little over two hours after his first killings he entered another student residence, Norris Hall, carrying several chains, locks, a hammer, a knife, two handguns with nineteen 10- and 15-round magazines, and nearly four hundred rounds of ammunition. With these he killed a further thirty-one people and wounded twenty-five others. Lasting two and a half hours, the Virginia Tech campus killings are the third deadliest civilian shootings in US history. The massacre was a living nightmare for the students, teachers and their families.

Anthony Mancini, who heads the Trauma, Social Processes, and Resilience lab at Pace University in New York, has been trying to understand how people cope with – and successfully adapt to – significant life stressors. Through his research, he observed that in most studies four prototypical patterns in how people respond to acute stress have been reported: resilience, gradual recovery, delayed reactions and chronic distress. Common to all four of these reported patterns is that people either manage to return to normal psychological functioning, or they get worse. What Mancini and his colleagues wondered, however, was whether there might be another pattern of responding: could traumatic events lead to psychological improvement for some people?

In a study published in *Clinical Psychological Science* in 2016,[1] he and his colleagues focused on 389 women who had been exposed to the Virginia Tech shootings. The volunteers completed a questionnaire two months after the shootings, and then again six months after the event. These women were selected from a larger sample that had already responded to a different survey before the shootings occurred, meaning the researchers could develop a rare insight into how this event changed people compared to their usual level of psychological functioning.

Mancini and his colleagues measured symptoms related to anxiety and depression, the level of exposure to the shooting, and perceptions of how threatening the shootings were. They also asked the volunteers about their level of social support as well as the extent to which they had gained or lost interpersonal resources, such as intimacy with family members and time with loved ones.

The researchers found several patterns in how different people responded to the event. Reflecting past work, they found that some people displayed low levels of anxiety and depression both before and after the shootings, consistent with a resilient response. Others showed increased distress immediately after the shootings, which then resolved over time, and others showed continued high levels of distress after the event. And then there was a new pattern of responding, one that had been largely overlooked or discounted as random error in prior studies. This pattern revealed that some participants in fact showed substantial improvements. These individuals had elevated levels of depression and anxiety *before* the shooting and experienced a marked reduction *after* it. These reductions were sustained for a year after the event, suggesting long-term benefits.

For several years, I have maintained a small, part-time psychology practice. It is always fascinating to see how some of the ideas and studies you read about play out in reality. I recall one client I will call

Steven, who was seeking help for depression. He had been to see me on a few occasions and this was not the first time he had struggled with such feelings. There were several apparent explanations for his current emotional difficulties and we were working through each of these, making some progress along the way, when one day he informed me his father had passed away. My immediate reaction was, 'Oh no, this is going to destroy him.' Steven had been struggling to cope with the many issues already on his plate, and was doing about the best he could. I just could not imagine how he was going to cope with another serious setback. About two weeks later Steven returned. He had travelled hundreds of miles to his dad's funeral. As he recalled the events of the funeral, including his contact with family members with whom his relations had been strained for many years, he exuded a certain calmness, even groundedness. He was more reflective and even-handed in his sentiments than he had been during his entire time in therapy with me. When I asked how he was feeling he said he was in fact feeling a lot better. Indeed, it appeared to me that his outlook had improved significantly.

Other research has found that, beyond having the capacity to alleviate emotional suffering, being reminded of death can also build positive character strengths. Two researchers from the University of Pennsylvania had a large-scale survey running on the internet in 2001.[2] Fortuitously (if one can use such a term in this context), this survey happened to extend over the period of the 9/11 terrorist attacks. Within the survey were questions intended to measure character strengths, including the amount of gratitude, hope, optimism, love and intimacy that people had in their lives, and how much they responded to others with kindness, were good at leadership, and worked effectively within teams. The scientists took data from more than four thousand respondents who had completed the survey around the time of the 9/11 attacks. By comparing those respondents who had taken the survey prior to the attacks to those who took the survey around one month after them, they found the above character strengths had increased. People were reporting more gratitude, hope, love and kindness and, what's more, these increases lasted for up to ten months after the attacks.

Together the evidence shows us that although traumatic events have the capacity to disturb healthy psychological functioning, they can also sometimes lead to improvements. There may be several reasons for this, but one Mancini and his colleagues noted was that tragic events, like the Virginia Tech campus shootings, can be a powerful trigger for the formation, and strengthening, of social bonds (a topic covered more in Chapter 5). Mancini found that the people who improved in response to the Virginia Tech shootings had low levels of social support initially. In response to the shootings, however, their levels of social support increased significantly and this mirrored their improvements in anxiety and depression. They got better, in part, because the tragedy of the shootings brought people closer together and built a stronger community around those who had experienced trauma. For some, this was a literal improvement on the levels of social support they had in their lives previously and led to improvements in their well-being.

There is another explanation. Painful experiences can act as a trigger for us to reorder our priorities in life and provide us with a new perspective. As noted by David B. Morris, Professor of Literature at the University of Virginia, in his award-winning book *The Culture of Pain*, 'Sometimes pain can even reveal to us beliefs and values we did not know we held . . . [It] can reorder priorities in a hurry. It can show us what truly matters.'[3]

The experience of pain – whether it be physical or emotional – puts us in touch with our sense of self in a way that we are not aware of when we experience pleasure alone. As Elaine Scarry, Professor of English and American Literature at Harvard University and the author of *The Body in Pain: The Making and Unmaking of the World*,[4] argues, it has the capacity to deconstruct us in a way that allows for new beginnings, and new understandings. It can break us down, sometimes in ways that can overwhelm and confuse us, but out of that state new meanings and identities can emerge, and we may be opened up to new possibilities and new horizons. Not only can our suffering create new meaning, but it can also remind us of what *is* meaningful (something discussed in greater depth in Chapter 7). As my patient Steven experienced, the pain of his father's death deconstructed his world, allowing him to rebuild it in a way that gave it new meaning, and ultimately also allowed him to feel less depressed.

Being exposed to the stresses and challenges of life can foster social connection and force us to re-order our priorities, but there is another even more important pathway through which adversity can build resilience: it literally shows us what we are capable of. When we confront difficult circumstances this provides an important source of information. By exposing ourselves to challenges, or even by being exposed to stresses we do not choose, we have an opportunity to experience our limits, to see how far we can go and what we can endure. This is not dissimilar to the notion of peak experiences discussed in Chapter 1. These events are also important sources of information. They can inform us of the limits of our personal resources, and when we survive such experiences we feel more confident that we can do it again.

It is exactly this kind of 'toughening' that soldiers around the world are exposed to as an ordinary part of their training. They may be required to hike all night with heavy packs, to endure long periods without sleep, or to overcome situations where they may fear for their own lives. They go through this training, not only because these activities build physical fitness, but because by surviving these experiences soldiers become more confident in their own capabilities and more prepared to meet the challenges of warfare.

Overcoming adversity demonstrates that we are competent, reminding us that we can achieve a level of mastery over something that is objectively difficult. Maintaining composure in the face of adversity is a clear demonstration of self-mastery and determination, leading people to experience that event as identity-affirming. Facing difficult experiences provides this test of character in a way that more pleasurable activities could simply never achieve – overcoming the pleasure of a foot massage is hardly a test of character! Through this process, pain reveals other qualities still, such as our ability to be independent and in charge of our own lives. When we are unable to handle challenges in life we are dependent on those around us for support. When we can overcome these challenges on our own, however, this reveals we are not dependent on others – we are the captains of our own ships.

Consider the examples of Laura Dekker and Jessica Watson, two girls who, separately, sailed solo around the world at the age of sixteen. The size of this challenge was evident in how people responded to the prospect of letting teenagers undertake such a massive adventure. Both girls and their families were cast as irresponsible for even attempting such a challenge, yet each was welcomed as a hero on their return. Jessica Watson was awarded Young Australian of the Year and the Medal of the Order of Australia. It is hard to even comprehend what that kind of solitude must feel like or the fear that must arise when hit by storms many kilometres from land, all alone. But it is exactly these elements that both young women overcame, and in so doing their adventures have no doubt led to a great deal of confidence in their own personal ability and self-worth.

Painful and stressful experiences are also critically important for our ability to satisfy our higher-order needs. In 1943 Abraham Maslow developed a theory of human motivation that is now one of the best-known theories within psychology.[5] He outlined a hierarchy of human needs extending from the most basic physiological needs (e.g. food and water) and the need to feel safe, to higher-order needs such as esteem and self-actualization. In describing self-actualization Maslow wrote, 'What a man can be, he must be.'[6] To achieve self-actualization we need to not only know what we *can* be, but have the opportunity *to be* all that we can be. This suggests we not only need to know what our limits are, but we need to extend and expand ourselves to meet those limits.

To understand this a little better, consider the ways in which we test new systems. Whether it is an engine or a cyber-security system, we do not know what these things are capable of until they are 'stress-tested'. Engines are run at high speed to determine what they can withstand, and people employ professional hackers to determine the strength of their cyber-security systems. If you are not into cars or computer systems, then you might have seen the IKEA machine that tests its chairs. In some stores, they have a machine repeatedly pushing down on a chair and releasing, endlessly, in order to demonstrate to the company's customers that their chairs are of high quality. We simply cannot know the qualities of something, what the full scope of its capabilities are, until we push it to its limits. When it comes to people, the story is the same – it is only through our adversities that we can experience all that we can be. It is our challenges and failures in life, it is our hard-won successes rather than our easy wins that reveal our capabilities. These experiences, in turn, build confidence, esteem and strength of character. Self-actualized people are better able to face the challenges of the world.

Evidence for the role of stressful experiences in promoting resilience or toughening is growing. Early research has found that rats shocked in infancy are less fearful as adults and are more active and adventurous in unfamiliar territory.[7] In the same way, during the Great Depression, children of lower-class backgrounds were found to be less fearful when surveyed in a daycare centre than children of upper-middle-class backgrounds.[8] Given the period, one might presume that the lower-class children would have experienced more painful and stressful events than those from relatively more privileged backgrounds. More recent research has begun to uncover a personality trait referred to as *hardiness*, which helps to buffer exposure to extreme stress.[9] Hardiness is defined by a commitment to finding meaningful purpose in life, the belief that a person can have control over the situations they are exposed to, and the belief they can grow and learn from both positive and negative life experiences. Hardy individuals are more likely to appraise potentially stressful situations as challenging rather than threatening, and are also more confident and better able to use active coping strategies and social support.

Stressful life events may even serve to benefit relationships, making them hardier and more robust. A study conducted by Lisa Neff and Elizabeth Broady from the University of Texas at Austin found that manageable stressors early in a couple's marriage can serve to make the relationship more resilient to future stress.[10] Over two and a half years, newlywed couples provided data regarding their stressful life events, their coping strategies and their marital satisfaction. Those couples who had good coping strategies and experienced moderate amounts of stress during the early months of marriage were not only better at dealing with stress later in their marriage, but also reported more satisfaction with their marriage. This was compared to couples who had good coping strategies, but who did not have to deal with any stress early on. Practice really does make perfect.

EVERYTHING IN MODERATION

Mark Seery from the University of Buffalo in the United States has also been interested in the link between adverse experiences and resilience or well-being. Specifically, he has been investigating how people's experience of challenging life events can contribute to developing a propensity for future resilience and improved psychological functioning.

To investigate this, Seery and his team developed a measure of cumulative lifetime adversity.[11] The questionnaire asks people to report how many adverse events they had experienced during the course of their lifetime. The number and frequency of different adverse events are then combined to create an overall score for lifetime adversity. In all there are thirty-seven experiences, ranging from the death of a loved one to experiencing a significant injury to sexual abuse. It is fair to say these events are more significant than placing your hand in an ice-bucket – yet after completing the survey this is exactly what Seery had his volunteers do next.

After completing the survey, Seery and his colleagues asked their volunteers to put their hand in a bucket of ice-water for as long as they could. They were then asked to rate their experience of the task on a measure assessing the extent to which they catastrophized the painful experience (e.g. 'I thought the pain might overwhelm me'). They also rated how unpleasant and how intense their experience of pain was and their current emotional state. What Seery did *not* find was that the more adversity a person experienced the better they were at coping with pain. He also did *not* find that no adversity produced better coping. Rather, he found a U-shaped relationship between how people responded to the ice-bucket task and the amount of lifetime adversity they reported. This can be seen in Figure 6 below. Those who had a moderate amount of lifetime adversity were least likely to catastrophize their experience, reported less pain intensity and unpleasantness, and less negative emotion in response to the ice-bucket. Those who had a moderate number of adverse experiences held their hand in the ice-water just as long as those who had many adverse experiences but were less able to cope with the experience. Those who had very few adverse experiences did not persist as long in the ice-bucket, but still reported more pain, catastrophized their experience more, and reported more negative emotions.

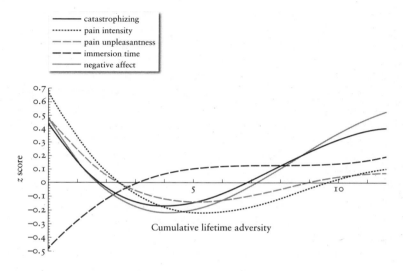

Figure 6. The dark dotted line (immersion time) shows how long participants held their hand in the ice-bucket. People with higher scores on the Cumulative Lifetime Adversity measure held their hand in for longer, although people with a moderate amount of lifetime adversity held their hand in for about the same amount of time as people with a lot of lifetime adversity. The other lines relate to how people responded to the pain of the cold water (i.e. whether they catastrophized the experience, rated the pain as intense or unpleasant, and how much negative emotion they reported). This shows that people with a moderate amount of lifetime adversity responded with lower scores on all these measures (i.e. they coped better) compared to people with little lifetime adversity and people with a lot of lifetime adversity.

Seery and his colleagues ran a second experiment, aiming to conceptually replicate this same finding. In this experiment, instead of asking the volunteers to experience pain they had them do a computer task which the volunteers believed was an intelligence test. This meant they would want to do well, and would feel stressed if they thought they were doing badly – no one wants to appear dumb. The computer-based 'test' required the volunteers to navigate an obstacle course as quickly and accurately as possible. While they were doing this they were hooked up to equipment that was assessing their heart rate, ventricular contractility (the amount of force generated by the heart muscle), cardiac output and total peripheral resistance (used to calculate blood pressure and blood flow). They did this because past research has demonstrated that although feeling both *challenged* and *threatened* increases heart rate and ventricular contractility, feeling challenged is uniquely marked by higher cardiac output and lower total peripheral resistance compared to feeling threatened. The experiment allowed the researchers to determine, at the physiological level, the extent to which the volunteers responded to the 'intelligence test' by feeling *challenged* or feeling *threatened*. As in the first study, they found a U-shaped relationship between cumulative lifetime adversity and their index of challenge vs. threat. Volunteers who had a moderate amount of adversity in their lifetime were more likely to respond to the task by feeling challenged, while those who had many or few adverse experiences responded to it by feeling threatened.

The research of Seery and his colleagues convincingly demonstrates that adverse experiences themselves can be a source of resilience – an effect that we can see even in the context of acutely painful or stressful events. Like Anthony Mancini above, Seery and his colleagues did not just find people can be resilient to stress, but that some stress is necessary for resilience to emerge – we are unlikely to cope if we have had few difficulties in our lives.

Seery and his colleagues wanted to explore this relationship further. They were interested in whether adverse experiences might not only make people tougher, but could also make them happier. To examine this, they again used their measure of cumulative lifetime adversity, incorporating thirty-seven life experiences including death, injury and abuse (as described above), and examined whether it might show a similar U-shaped relationship with well-being.[12] The researchers reasoned that a moderate amount of lifetime adversity, as compared to no lifetime adversity or high levels of adversity, would be most likely to foster toughness and mastery of life, and this in turn would facilitate the highest levels of well-being. This reasoning is consistent with past research showing that overwhelming levels of adversity reduce one's ability to manage stress in life. This means that stressful experiences do not foster a sense of mastery, and in turn resilience to these experiences is never developed. On the other hand, being sheltered from adverse experiences in life temporarily protects against distress, but also provides no opportunity to develop toughness and mastery. Those who have been sheltered from adversity, or who have been overwhelmed by it, are less able to cope with the inevitable stresses and challenges that life presents. This in turn means that these individuals would be less likely to experience high levels of well-being.

To test these predictions, the team collected data from 2,398 adults living in America over four years, between 2001 and 2004. In the first of four separate surveys, the researchers administered the same lifetime adversity measure as above. Over the next three years they administered the second, third and fourth surveys, where they asked people to indicate whether they had experienced any of these same adverse events in the previous six months. This allowed them to assess how resilient their volunteers were to recent stressful life events. The volunteers also completed several measures designed to capture their overall levels of well-being, indicating their levels of overall distress, their overall satisfaction with their lives, and the extent to which their emotional or physical health interfered with their social or work activities (functional impairment).

In support of their predictions, and consistent with the above findings, the researchers observed that people who indicated a moderate number of adverse experiences (out of the list of thirty-seven included in their measure of cumulative lifetime adversity) were also those who reported less overall distress in their lives, greater overall satisfaction with their lives, and that their emotional or physical health was less likely to interfere in their lives (less functional impairment). In short, as predicted, it was the people who had experienced moderate adversity who also reported the highest levels of general well-being – they were the happiest.

The researchers also found that the more people had experienced recent adversity over the previous six months (as measured in the second, third and fourth surveys), the higher levels of global distress and more functional impairment they reported; overall they were less satisfied with how their lives were going. Yet, again, it was the people with a moderate amount of lifetime adversity that were the least affected by these recent events.

You might have noticed the study began in 2001. As with the University of Pennsylvania research mentioned earlier in this chapter, this allowed for an unexpected analysis of how people responded to the 9/11 attacks. This event was not only traumatic for those living in New York; in fact it is now widely accepted that people all over America experienced trauma related to these attacks. The extent of damage and loss of human life is largely unprecedented within recent American history. To capture responses to this event, the researchers also included a measure of post-traumatic stress-related symptoms, focusing specifically on how people responded to the attacks. Again they found the same pattern, a U-shaped relationship. Across America, it was those people who had experienced past traumas – be they accidents or injuries, illness, death, loss, divorce, physical or sexual abuse, financial stress, natural disasters, discrimination and prejudice or parental neglect – to a moderate degree over their lifetime who coped best.

This same pattern of trauma leading to improved psychological functioning led Friedrich Nietzsche to coin the now-famous phrase 'that which does not kill us makes us stronger'. It is also a recurrent theme in the work of many writers and thinkers in history. In Dante's *Divine Comedy*, he describes his search for his lost love, taking him through hell and purgatory, eventually reaching paradise. In *Crime and Punishment*, Fyodor Dostoevsky tells of Raskolnikov who is punished for his murderous crimes, but who gains redemption and moral regeneration when he embraces the suffering of the prison camps.

The idea that adversity can build resilience is not new; we know that sometimes people come out of tragedy stronger than they were before. What we seem to have forgotten, however, is that adversity is not a choice when it comes to well-being. If we want to be capable of coping with life and, in turn, feeling happy and satisfied, we need to face some adversity. Hiding from it leaves us vulnerable.

THE HOLY GRAIL

Not everyone who experiences adverse circumstances responds in positive ways. People torn apart by trauma can experience post-traumatic stress disorder (PTSD), a condition that is incredibly debilitating. Just think of the many returning servicemen and women who lose their lives to PTSD following the horrors of war. Yet, as the concept of post-traumatic growth suggests, some can also emerge from these experiences stronger than they were before, exhibiting positive changes that go beyond their previous levels of functioning and well-being.

If psychologists could pinpoint exactly what qualities or which psychological processes could predict when people will respond to trauma with resilience, rather than falling apart, it would be like finding the Holy Grail – it would tell us what makes humans psychologically strong. The fact is there is no silver bullet, but what we do know is that a very broad range of factors are important. These may include family functioning, social support, personality traits associated with high degrees of flexibility and low amounts of rigidity, emotional intelligence, and cognitive capacity. More informative, however, is that these broader background factors may feed into two more immediate and tangible indicators that can predict coping.

The first is related to whether people can create meaning or coherence in response to tragedy and trauma. This was an insight championed by psychiatrist and Holocaust survivor Viktor Frankl, who in 1946 wrote a book titled *Man's Search for Meaning*. It has sold over 10 million copies and been listed as one of the ten most influential books in the United States. In this, and in his other books, Frankl wrestled with his experiences in the Nazi concentration camps and what it was that allowed some to survive and overcome these experiences, while others perished. For Frankl, it was about finding meaning and purpose.

He recounts an experience when he and his fellow prisoners were being driven by guards over stones and ice along a dark road leading from the camp. During this extreme suffering his mind turned to his wife and he recalled how much he loved her. At this moment, he was overcome with a sudden insight into the nature of love – that it is the ultimate and highest goal to which man can aspire. He tells how he felt he had understood something terribly significant. He realized that even when a man has nothing left, when he is in the midst of terrible suffering, he can find bliss in the contemplation of those whom he loves.

When people can establish purpose in their lives, and develop a sense of autonomy, mastery and self-acceptance, they are better able to cope with challenges. This is what the late Aaron Antonovsky, an Israeli-American sociologist, referred to as a 'sense of coherence'. He argued that when people can develop coherence in their lives this promotes both mental and physical health and well-being.[13]

A second factor that determines when people respond to trauma with resilience has to do with how people interpret their adverse experiences. A well-established theory within psychology, developed by Jim Blascovich and Joe Tomaka,[14] details how framing a stressful event as a challenge, rather than as a threat, fundamentally changes how people respond to that event. This approach was used by Seery and his colleagues in their research detailed above, and as in that work, these different framings have been linked to distinct under-lying physiological reactions. According to this theory, what determines whether a particular event is framed as challenging or threatening is whether people feel they have the personal resources to cope. Take, for instance, public speaking. Many people may find this threatening when they first try delivering a talk to a room full of people, esp-ecially if they feel tired, or perhaps under-prepared. However, if they feel they have sufficient resources – they are well prepared, have had a good night's sleep, and have practised before – they are more likely to find it challenging. Put simply, people experience stressful events as a *threat* when the demands of coping with the event are greater than their personal resources to cope with those demands. However, when their personal resources are thought to be greater than those demanded by the situation, people will experience the event as a challenge.

When people respond to adversity by embracing it, framing it as a challenge and finding a sense of coherence or even purpose within it, they are more resilient. Yet, as with Seery's work above, this capacity to see challenges rather than threats, and to find coherence, does not come without practice. There is an important cycle in which exposing ourselves to adversity in life builds our capacity to respond with resilience – to see the event as challenging and meaningful – and this in turn ensures we are more likely to seek out such experiences, which in turn continues to build resilience and reap rewards. For if you don't try, you can't fail.

Although we are only just beginning to understand processes that promote resilience in individuals one thing is clear: *exposure* is a necessary component.

BUILDING BIOLOGICAL TOUGHNESS

Richard Dienstbier at the University of Nebraska–Lincoln developed what is referred to as 'The Toughness Model' in a paper published in *Psychological Review* in 1989.[15] He begins with the observation that exposure to stressful or painful challenges increases physiological arousal. Arousal effectively consists of the release of increased amounts of adrenalin, noradrenaline, dopamine and cortisol into the brain, which in turn intensifies our levels of alertness and preparedness for action. It is exactly this process which increases our heart rate and gives us sweaty palms before giving a speech or jumping out of a plane.

He notes that when researchers have exposed young rats to a range of stressful situations, extending from electric shocks to extensive handling or living with an 'aunt' of a different species, these rats developed bigger adrenal glands. The adrenal gland is responsible for the release of adrenalin, noradrenaline and dopamine into the brain and, as noted above, these chemicals lead to increased arousal. (Although it is not strictly accurate, I am just going to refer to all of these as adrenalin from now on as this is the more familiar term.) By developing bigger adrenal glands these rats had developed an increased capacity to release adrenalin and as such one might have expected that they would show higher levels of 'arousal' in response to stress (that is, a greater emotional response to stress). Yet, this was not the case. Rather, these rats were calmer, less emotional and less fearful when exposed to threats such as an electric shock. In fact, they appeared to have lower levels of adrenalin in their system when just resting in their cage (this is often called a lower base rate, a bit like marathon runners have lower resting heart rates), but when exposed to a stressful situation their adrenal glands could respond more effectively by releasing large amounts of adrenalin. They also showed a faster return to their low-arousal (low adrenalin) state, meaning that they reacted to threats quickly and efficiently, but did not remain stressed for long afterwards.

These rats not only had lower resting levels of adrenalin, they also had lower resting levels of cortisol. Furthermore, their increased capacity to release adrenalin in response to stress did not extend to an increased tendency to release cortisol. Like adrenalin, cortisol is released into the brain in response to stress and acts to increase arousal and suppress inflammation, but if it is kept at elevated levels for long cortisol can have negative effects on the body, such as leading to a depressed immune system, muscle-tissue wastage and reduced brain function. High base rates of cortisol have also been linked to depression, anxiety and anorexia. Unlike adrenalin, cortisol undermines rather than facilitates performance. All of this is to say that while the rats in the above study had developed a more efficient and effective response to stress, this response was specific to adrenalin rather than cortisol. They coped with stress in more adaptive and less maladaptive ways.

Martin Seligman, the leader of the positive psychology movement and one of the most famous psychologists in the world today, was also interested in rats. In fact, his earliest studies on learned helplessness showed that when rats had no control over whether they would experience a stressful event, such as an electric shock, they would demonstrate 'helplessness' when faced with another shock soon afterwards. That is, they would stop trying to escape and just endure the shocks with few attempts at avoidance.[16] This research provided major insights into the nature of depression and other disorders where people appear to give up in life. Relevant for us, however, is this research was later extended by a team led by Jay Wiess, from the Emory University School of Medicine, to show that the rats who responded with 'helplessness' suffered from reduced adrenalin levels in the brain. The researchers determined that the prior stressful experience had depleted the rats' capacity to produce enough adrenalin to respond appropriately to the second stressful event. This in turn meant they experienced low levels of arousal, even when faced with an electric shock.

The researchers conducted a follow-up study where instead of exposing the rats to severe levels of prolonged stress – as in the learned helplessness studies – they tried exposing them repeatedly to stress, but for shorter periods and allowing time for sufficient recovery in between.[17] For fourteen consecutive days, the rats were trained by either swimming in cold water or experiencing electric shocks for short episodes, followed by recovery periods. The researchers found that when the rats were exposed just once to the electric shock or cold swim they showed the 'helpless' response to a second stressful task. In this case the second task involved the rats being placed in a box where they would hear a noise that was followed by a shock. They could escape the shock by jumping over a barrier when they heard the noise, but if they did not they would get shocked. This task was repeated multiple times so they could learn that the noise was a warning the shock was about to happen. The rats who had experienced just one episode of cold-water swimming or electric shock tended to sit still and let the shock happen to them – this was the learned helplessness response. Interestingly, however, the rats that were exposed repeatedly over fourteen days to the cold swim or electric shock no longer showed this response. Instead, they escaped more quickly and efficiently, avoiding the second shock rather than sitting helpless and letting it happen.

Evidence that the ability to efficiently release adrenalin in response to stress, while also maintaining a low base rate of arousal (like our rats above), is good for performance has been demonstrated in a number of studies on humans. For instance, Swedish schoolchildren who showed lower base rates of adrenalin, but a larger increase in adrenalin levels when exposed to a stressful situation (e.g. completing a maths test), performed better.[18] A similar finding was reported when researchers examined Norwegian Army paratroopers, finding that a greater increase in adrenalin levels from before a parachute jump to after the jump was associated with better performance – not only in jumping out of planes, but also in their competence in technical writing tasks.[19]

Beyond performance, this ability to rapidly increase arousal (by releasing more adrenalin into the brain) but to maintain a low base rate of arousal when resting is also associated with a range of positive personal traits. For instance, the Swedish students who showed this same pattern of arousal were also more satisfied with school and had better social adjustment and emotional stability, according to their teachers. Other research has found this pattern of responding is associated with increased self-reported stress tolerance and lower levels of neuroticism (a personality trait associated with emotional instability).

To bring this back to our discussion above, we know that when people feel they have the resources to cope with a stressful situation they experience this as a challenge rather than a threat. Dienstbier's theory provides us with a physiological mechanism for understanding how this might work. Through the process of toughening we improve our ability to respond more effectively to stress – via the more efficient release of adrenalin into the brain – meaning we literally have more energy and therefore more personal resources to cope. As we become aware of the availability of these increased resources we are more likely to confront novel stresses as challenging rather than threatening.

To take this a step further, research has found that when people feel *challenged* by a task it increases the release of adrenalin, but reduces how much cortisol is released. When people feel *threatened* and as though they are unable to cope, they release more cortisol and, if prolonged, these threats can also deplete the levels of adrenalin released into the brain. Recall that increased adrenalin is associated with better performance and increased cortisol is associated with poorer performance. Feeling challenged will mean we perform better than when we feel threatened, and this can be explained at the physiological level. This in turn assures we exit these challenges feeling we have mastered the situation, rather than been overwhelmed by it, prompting us to seek out challenging and therefore toughening experiences in a cycle resulting in increasing toughness and enhanced emotional stability.

So how can we make use of this information? Well, Dienstbier notes that we can engage in 'passive' or 'active' toughening. Passive toughening effectively means being exposed to situations that facilitate a certain level of 'stress'. Like the rats above, exposing ourselves to the cold of a swim in the sea during winter would have similar effects. Active toughening could be achieved by engaging in behaviours such as regular aerobic exercise. His theory suggests that these kinds of physical experience should toughen us in ways that mean we are likely to become more emotionally stable, and less susceptible to anxiety and depression. As I noted in Chapter 3, there is plenty of evidence that running or even weightlifting can serve to alleviate depression. Dienstbier's theory suggests another pathway through which this may work. Intensive and regular physical activity or exposure to stressful but challenging experiences will not only alleviate depression or anxiety, but can toughen us so that we are less likely to experience them in the first place.

STRESS REDUCTION IS NOT ALWAYS THE BEST THERAPY

Biological systems become stronger when exposed to stress. Immunization is a good example (as mentioned in Chapter 2). By exposing the body to controlled doses of a given bacterium we trigger the immune system into building a more effective response. This in turn leads to a stronger immune defence system and a reduced likelihood of infection when future contact with that bacteria occurs. In effect the immune system responds to a moderate level of stress (immunization) by becoming stronger and developing better systems to cope with more significant levels of stress (potential infection) in the future.

When the biological system has not built sufficient defences, and infection takes hold, we are forced to turn to one of the greatest life-saving inventions in human history – antibiotics. But there is a downside; antibiotics strengthen bacteria, not the biological systems they were designed to protect. The more antibiotics we use, the more antibiotic-resistant strains of bacteria we create. Each bacterium that survives a dose of antibiotics adapts and becomes stronger. Furthermore, the more that we rely on antibiotics to kill these bacteria, the more we prevent our immune system from building its own defence systems against infection.[20] Indeed, in our attempts to kill bacteria we are creating 'superbugs', and have put humanity on the verge of the potential for an outbreak of untreatable disease.

Exposure to stress (such as pain and trauma) can be a catalyst for humans to develop a stronger and more resilient response to future stress; it can literally build 'psychological immunity'. What is important to consider, however, is that our current approach to psychological therapy is more akin to 'antibiotics' than 'immunization'. Unlike Dienstbier, we have tended to think of arousal as having negative consequences and it is perhaps for this reason that many of our psychological interventions have traditionally aimed at lowering arousal levels – just think of breathing techniques or meditation. This is because we have often considered arousal as an important factor not only in psychological disturbances such as anxiety disorders, but also physical illness. Studies have linked Type-A personality or neuroticism – two personality types defined by a tendency to experience high levels of arousal – to medical conditions such as hypertension or even cancer. On this basis, arousal does not appear to be a very good thing, but the concept of psychological immunization challenges this view.

It is worth briefly considering this link between stress and cancer, and how it might be linked to the differences between prolonged stress and short bursts of stress. Just as prolonged stress can lead to psychological helplessness, as reviewed above, it can also lead to poor immune function and disease – a kind of immune 'helplessness' if you like. Yet, as we have seen, short periods of stress can boost adaptive responding by enhancing rather than depleting adrenalin production, which, in turn, can strengthen the immune system and prevent disease. Support for this comes from a study where rats were exposed to intermittent (as opposed to prolonged) stress, which resulted in enhanced immune function and resistance to all stages of cancerous growth. Moreover, these effects are boosted if the rats are given some level of control over the stress – when they could experience it more like a challenge than a threat.[21]

To return to the concept of psychological immunization, it is interesting that Dienstbier notes the minimal impact that positive views of arousal have had on therapy. He noted this back in 1989, but not much has changed. Now, as then, psychological therapy is mostly focused on strategies that seek to reduce levels of arousal. From the toughness perspective, however, such therapies may prioritize the short-term benefits of low arousal at the expense of the long-term benefits of toughening – they may focus people on the avoidance of the very situations that can lead to toughness. To be fair, there are many approaches, such as desensitization therapy, that aim to expose people to stress or fear but in an environment where they feel safe and can cope. Yet, there remains little recognition of the intrinsic therapeutic benefits these experiences can offer.

This same drawback is evident in the case of medications such as beta-blockers. They reduce anxiety and arousal, but by doing so prevent people from building their own responses to stressful or painful situations. Whether we are using medication to ward off illness or to ward off stress, we are undermining the capacity of our natural biological and psychological systems to develop resilience – to become tough.

The key to healthy psychological functioning is exposure. If we want to be happy, we cannot afford to hide from our challenges and surround ourselves in protective layers of comfort. To achieve emotional stability and the capacity to handle challenges when they arise, we may be well advised to occasionally seek out discomfort and to take ourselves outside our proverbial comfort zones more often than we do.

5

Connecting With Others

My two six-year-olds love sticking plasters. The smallest scratch is cause for a plaster and will be referred to continuously until it is no more than a figment of their imagination. Scratches, bruises and bumps are excellent ways to get attention from adults, and children understand this early on. It does not take long for children to exploit the attention-grabbing aspect of getting hurt.

Humans have evolved to communicate their pain so that others experience an empathic response. For instance, humans have a distinct facial expression for pain (referring specifically to physical pain here), which is detectable from infancy to old age. No matter which kind of pain-eliciting experience people are exposed to, their expression is more or less consistent. What is most fascinating is that our expression of pain is more easily detected by observers when we try to hide it than when we intentionally express or exaggerate it.[1]

Just as the expression of our own pain is difficult to control, so too witnessing the pain of others can lead to a gut-level reaction – a response that can be observed in the brain. A team of researchers from the University of Washington showed volunteers a series of 128 photos of hands and feet.[2] In half of these photos the hands and feet were in a situation that would be likely to cause pain, while in the other half they were shown in very similar situations, but which would be non-painful. For instance, in one photo a person's finger would be shown as if it had just been jammed in a door, while in the comparison photo the hand would be shown on the door knob while shutting the door. The volunteers viewed these photos while they were in an fMRI scanner so the researchers could observe the patterns of brain activation associated with each photo. In order to elicit engagement with the photos, volunteers were asked to assess the level of pain experienced by the person shown. What the researchers found was that viewing and assessing pictures of people in pain activated the same brain regions triggered by a person's own painful experience. At the neural level, we literally mirror the pain of others, and this occurs automatically – our empathy is real.

This tendency to experience empathy for the pain of others is evident across a number of different kinds of negative experiences. Another group of researchers, from the University of California, examined whether this same neural empathic response for another person's pain may be evident when we observe others experiencing social pain (rejection or exclusion by others).[3] While the volunteers were in the fMRI scanner they watched the same game referred to in Chapter 3, where a person was socially excluded. This game is called Cyberball and is commonly used in the field of experimental social psychology to elicit the experience of social ostracism. Just to remind ourselves, people are told they are playing against two other players, when in fact the game is programmed to either include the volunteer or to exclude them. It is a surprisingly powerful experience, regardless of whether people realize it is just a computer program.[4] Even if volunteers are told their co-players are members of the Ku Klux Klan, they still feel upset by being excluded.[5] We are incredibly sensitive to social rejection; just like physical pain, social pain activates an evolved threat response. Imagine being out there on the savannah back at the beginning of human history and your survival depended on being protected by your tribe. Without them you would be very unlikely to live. Just like physical pain warns us of potential survival threats, so does social pain. This explains why the experience of being socially rejected literally 'hurts', even though in the modern world this survival threat is obviously diminished.

Returning to the fMRI scanner, the volunteers in this instance did not play the game themselves, but they watched what they believed was another person playing the game, against what they believed were two other real players. In fact, the researchers went to quite some length to maintain the plausibility of this – even stopping the experiment at one point and telling the volunteer it was because one of the players had to use the bathroom. The volunteers observed two rounds of the game being played. In one round the player was included, receiving the ball an equal number of times, while in the other round they were excluded, receiving the ball significantly less often. Next the volunteers were told they could email the players their feedback. Actual email accounts were created for this purpose, and they were told they could write whatever they wanted about what they had observed.

These emails were next rated by eighteen people who had no knowledge of the purpose of the study. Specifically, they were asked the following questions:

1. Does it seem like they are trying to comfort this person?
2. How supportive are they toward this person?
3. How much do they seem like they are trying to help this person?

This was in order to determine how 'pro-social' participants were in their emails. An especially pro-social email sent to the victim of social exclusion is given below:

> Dear Adam, while watching your game of Cyberball I noticed you may have felt left out when Erika and Danny were consistently throwing the ball to each other. I just wanted to say I'm sorry that happened and I am sure there is some explanation that has nothing to do with you. You seemed to be a great ball thrower.

The findings revealed, first, that when watching the person being excluded by the other players, compared to when they were included, the volunteers showed greater neural activation in areas associated with mentalizing – that is, the tendency to take the perspective of another person. Most interesting was that when the researchers looked at the association between this pattern of brain activation and the level of empathy expressed in the emails, they found a directly positive relationship. The more watching others being rejected activated empathy-related activity in the brain, the more likely people were to write empathic emails to the victim.

Whether it be physical pain or social rejection, we vicariously (and to a large extent automatically) experience other people's feelings and, in turn, respond to them with pro-social behaviour. These negative experiences literally draw people together in ways that are often beyond conscious control.

EXPLOITING THE LURE OF PAIN

Although pain leads us to communicate and respond to others automatically, there are also plenty of occasions when people seek to exaggerate their negative experiences. Whether it be physical, social or emotional pain, people can still focus on and exaggerate the effects of those experiences. According to researchers and clinicians, these people are employing the same strategy as my six-year-olds: this motivation to exaggerate one's pain to attract social support is referred to as 'secondary gain'. In these cases, a person's expression of their experience goes beyond the motivation to elicit assistance to deal with their struggles (referred to as 'primary gain') and extends to the satisfaction of other motives such as attention and concern from others more generally.

An especially prominent example within Australia of exactly this type of behaviour is Belle Gibson, an app developer and blogger who attracted a great deal of attention by claiming that she had cancer. She claimed it had entered her liver, kidneys, bloodstream, spleen, brain and uterus. She also claimed that she had undergone heart surgery several times and had momentarily died on the operating table – and that she had suffered a stroke, just in case the rest wasn't enough! It was on this basis that she also managed to get her healthy-eating book published, her related food app planned for inclusion on the first Apple Watch, and a significant internet following. As it turned out, she had experienced none of these conditions. However, by claiming to have experienced great suffering Belle Gibson built a significant business, thanks to people's empathy and admiration at her ability to overcome it.

Although not everyone goes to quite the same lengths as Belle Gibson to get attention, there is a general tendency, observed by medics and psychologists, for people to 'catastrophize' their suffering, to respond to anticipated or actual pain with increased negative cognition and emotion, for the same reasons. Research has demonstrated that pain catastrophizers tend to exaggerate their hurt in order to maximize the probability of recognition. This may include exaggerating facial expressions or vocalizations. Importantly, researchers have found that pain catastrophizers tend to engage in these exaggerated responses only when other people are present.[6] Other research has revealed that people who feel anxious about their social relationships are more likely to catastrophize their pain, strongly suggesting that exaggerating pain is used to secure attention and empathy from others.

One particularly illuminating study examined this side of pain in the context of guilt and punishment.[7] Two experimental psychologists, Kurt Gray and Dan Wegner, had volunteers come into a laboratory and told them they would be listening to the recordings of another participant, whom the researchers referred to as Carole. The volunteers were then told that Carole thought the study was about 'chance and winning' when in fact it was about moral behaviour – in reality it was about neither. They were also told Carole would roll an eight-sided dice and that she believed she and her partner in the study would receive some money depending on the outcome. Furthermore, that Carole believed if she rolled an eight she would receive $5.50 while her partner would receive nothing and that, since no one was actually watching the dice-roll, Carole might have been tempted to cheat to win more money for herself. Sure enough, the volunteers were informed Carole had reported rolling an eight, winning more money for herself. Next, they were told that although there was no way to know if Carole had lied or had legitimately rolled an eight, people often admit to wrongdoing when placed in stressful situations. To this end, they were led to believe that Carole had been asked to perform a painful task – putting her hand in a bucket of ice-cold water for 80 seconds. Volunteers then listened to a recording of Carole being 'tortured' and could judge whether or not they thought she had cheated. Half of the volunteers heard Carole react to the painful task with little discomfort, reacting stoically to the cold. The other half of the participants heard Carole reacting with significant discomfort, whimpering throughout the painful task. The volunteers then evaluated how likely it was that she had cheated or was lying, and how immoral they considered Carole to be. Results showed that the volunteers who heard Carole expressing a high level of pain thought she was less guilty compared to the volunteers who heard Carole express-ing a low level of pain. The more pain Carole appeared to be experiencing, the less guilty she was judged to be, further supporting the idea that expressing pain communicates the need for protection and support, with the potent side effect of reducing others' motiv-ation for blame and retribution.

This suggests innocent people who opt to keep an unemotional demeanour when being tried for alleged crimes in a court of law are doing themselves no favours. A famous case in Australia was that of Lindy Chamberlain, convicted in 1982 of murdering her own baby, although she claimed it was the work of a dingo. The standout feature of her trial was her choice to show no emotion, no pain. In the end this was her undoing: she was convicted and given a life sentence, spending four years in prison until the child's jacket, found near a dingo's lair in 1986, proved her innocence.

PAIN MOTIVATES CONNECTION

Pain can draw others to us, but it also works the other way as well; when we experience pain ourselves it motivates us to draw close to others. It literally brings people together.

One reason pain increases our motivation to affiliate with others is that social support is a very effective antidote to pain. When your child bumps their knee, a warm cuddle is all they need. In the same way, when we are depressed, afraid or stressed, other people are sometimes the best medicine.

The effect of social support in reducing pain was investigated by a group of researchers at the University of California.[8] They recruited people in heterosexual relationships for the study and took one of the partners (the male) into a separate room to have his photograph taken for later use. They then put a hot probe against the arm of their female partners, first to determine their individual pain thresholds. The researchers varied the amount of pain for each volunteer until they rated it as a 'ten' (out of twenty), indicating a moderate amount of pain. Next, these same volunteers received a total of eighty-four thermal stimulations to their arm, which was placed behind a curtain so they could not see what was happening. They were led to believe the stimulations could be either more or less painful, but in fact they were the same each time.

The stimulations were spread across six different conditions. Sometimes the female volunteers were holding the hand of their partner (as he sat behind the curtain). Other times they held the hand of a male stranger (also behind the curtain), or they held a squeeze ball. In other conditions, they viewed the photographs taken of their partner, photographs of a stranger, or photographs of an object. In one condition, they simply viewed a blank screen with a cross on it. This was to provide a baseline against which to compare the effectiveness of the photographs and hand-holding in reducing pain.

During each stimulation, the volunteers pointed to a number on a 21-point scale to indicate how 'unpleasant' the stimulation was. The experimenters also wanted to rule out one possible alternative explanation which was unrelated to their predictions about the role of social support: that having one's partner present, or viewing a photograph of your partner, is simply more distracting so that pain is less apparent. This would mean that social support itself is not especially effective for pain relief, but just any distraction would do. To rule out this possibility, the researchers had participants perform a reaction-time task where they pressed the space bar on a computer keyboard in front of them when they heard a beep. This occurred randomly throughout the experiment, allowing the researchers to determine if the volunteers were slower (and by inference more distracted) in the social support conditions. The results confirmed the researchers' predictions. Participants under the various conditions did not differ in how distracted they were (measured by response times to the beeps). However, holding a partner's hand led to significantly lower pain ratings compared to holding an object or holding a stranger's hand. This same pattern of results was evident when viewing the photographs. Viewing the photograph of their partner reduced the pain ratings of the female volunteers compared to when they viewed photographs of an object or photographs of a stranger. Feeling emotionally connected dulled feelings of pain.

Social support has pain-attenuating effects. This outcome is not felt with just anyone, but rather those we know and trust to offer us social support. Somewhat surprisingly, this effect of a loved one's presence can even be triggered through simply viewing a photograph. It is no wonder then that we often keep photos of those we love on our phones, or in prominent locations such as a screensaver.

If social support is so effective in reducing pain, do people actively seek out the salve of such support when they are in pain? With so many studies showing social support as effective, my colleagues and I wondered whether people who experience even just mild transient pain in the laboratory might be motivated to seek out the company of others. We considered this would be a significant finding, as it would show even mild and non-injurious pain can promote affiliative behaviour, and, of course, affiliative behaviour has the potential to build long-term supportive relationships, and in turn social benefits.

We knew of one study that had come close to these ideas, conducted by the psychologist Stanley Schachter at the University of Minnesota in 1959.[9] He invited college women to participate in an experiment that would involve experiencing electric shocks, but split them into two different groups. One group of participants were told that the shocks would be severe (they were labelled the high-anxiety group); the other group were told the shocks would be mild and nearly painless (the low-anxiety group). While they were waiting, the women were asked to rate their anxiety, and also to indicate if they would prefer to wait alone or with others. What Schachter found was that 63 per cent of the women in the high-anxiety group preferred to wait together, while only 33 per cent of the low-anxiety group expressed the same desire.

This demonstrated that fear of pain – a negative experience in itself – increased the motivation to affiliate. In a follow-up study, Schachter took this finding one step further. He again made people feel anxious about impending electric shocks, except this time he gave them the choice to wait with others who were also going to receive shocks, or to wait with others who, he told the volunteers, were going to speak to an academic adviser. Providing the first empirical evidence for the phrase 'misery loves company', Schachter found that the women who were anxious about being shocked preferred to wait with others who were going to receive electric shocks as well. Not only do we want to associate with others when we are feeling anxious, but we want to associate with others who share our feelings of anxiety.

While the 'misery loves company' effect has become part of common wisdom, the reasons for it are slightly counterintuitive – it might seem more logical to seek out someone who is calm and therefore better able to care for our needs. One reason we seek out anxious others instead could be that we like to be around others who know what we are going through, and this sense of shared experience is especially effective in helping us to cope. As you will see in the next section, it may also be because sharing painful experiences with others promotes a particularly powerful form of affiliation – it leads to feelings of bonding and solidarity.

First, let us return to this basic effect of increased affiliation. The Schachter study showed that when people are anticipating something threatening, they prefer to 'huddle' together, allowing them to draw on social support from others who are subjected to the same fate. The question remains, however, as to what would happen after those shocks. Would those same people still want to affiliate with others?

This is what my colleagues and I set out to measure in our own laboratory.[10] We invited university students to come and participate in a psychological experiment, and we told them that it was a two-part study looking at physical acuity and interpersonal interactions. One group of participants – the pain group – did a physical task that required them to locate metal balls in a bucket of ice-water. They did so until they could no longer tolerate the cold. The other group – the pain-free group – did exactly the same task, but with room-temperature water in the bucket. They were stopped after 90 seconds. The experimenter then told each of the participants they were to have an interaction with another person who was currently completing a different study in a different room. To make this seem plausible, we had left a chair at one end of the room with a jacket and bag on it (which were as gender neutral as possible). We then asked our participants to get ready by pulling their own chair over in front of the other chair with the other participant's belongings on it. After they moved into position we then gave them a clipboard with some questions for them to answer.

In the study our volunteers never met any other participant; rather they were simply debriefed and allowed to leave. What we wanted was to measure the distance between where they had placed their chair and the chair which they believed the other participant would be sitting on. This chair-distance measure is commonly used in psychology experiments as an indicator of how close people want to be to others. For instance, it is commonly used as a measure of prejudice. People who dislike members of other groups tend literally to 'keep their distance'. For our purposes, we used this as a measure of affiliation; we wanted to know whether our volunteers who experienced pain would draw their chair closer to that of the other participant.

Confirming our predictions, we found that participants who had experienced pain pulled their chair closer than the participants in the control group (room-temperature water). This showed that just as Schatcher's volunteers had sought to affiliate with others when they were anxious about receiving electric shocks, our volunteers did the same thing, but this occurred after the actual pain had passed. To be clear, our volunteers did not experience a significant threat, just a painful cold feeling on their hand. Yet, this was sufficient to increase their motivation to seek out social connection. This has a range of practical implications and suggests that even everyday common experiences of discomfort may lead to affiliative behaviour.

BEYOND COPING

What we have covered so far shows that when we stub our toe or feel sad we reach out for others because they make us feel better. Yet, the social side of such painful experiences is not limited to coping. Although this might provide the primary motivation to seek out connection with others, our pain can also do more. Beyond bringing us together, our negative experiences can also foster specific qualities conducive to positive interpersonal relationships.

A good example of this comes from the work of Bernadette von Dawans and colleagues from the University of Freiburg in Germany. In a study published in 2012,[11] this team of researchers investigated the effects of social stress on pro-social behaviour. They had male students come into the laboratory and seated them in a room where they were not allowed to communicate with each other. The volunteers were then provided with a heart-rate monitor and the researchers measured their baseline levels of cortisol by taking saliva samples. The volunteers were then split into groups. Half of the volunteers were recruited simply as interaction partners and were not exposed to any stress.

The rest of the volunteers were administered the Trier Social Stress Test for Groups. To understand the nature of this test you just have to imagine the worst possible job interview. In the 'stress' condition of this test, participants are told they need to prepare for a job interview of their choosing. They have 10 minutes to prepare for this, and then they have a 12-minute interview with a panel of individuals who represent the 'evaluation committee'. They are informed that, as part of the interview, they will need to deliver a 2-minute speech trying to convince the committee as to why they are the best person for the job. The evaluation committee are trained to withhold verbal and non-verbal feedback and the volunteers are informed that the committee members are in fact experts in the evaluation of non-verbal behaviour. Together these elements are designed to put people into a context of high social evaluative stress.

If all of this is not enough, the volunteers have to complete this task with the other volunteers watching on, and if they do not complete their 2-minute speech the committee members simply say, 'You still have some time left. Please continue!' If the volunteers finish a second time before their 2 minutes are up, the committee members stay quiet for a full 20 seconds before asking their prepared standard questions.

After these 12 minutes of 'the worst job interview you have ever experienced', the volunteers were then asked to serially subtract the number 16 from a given number (e.g. 4,878, 4,862, etc.). If they made a mistake a member of the evaluation committee would say 'Stop. Please start again.' This part lasted for a full 8 minutes.

In the control version of this test, the volunteers read a popular scientific text for 10 minutes and were explicitly told that their reading performance was not being evaluated. They were then asked to read the text out loud in a low voice along with other participants for 12 minutes, and finally to enumerate series of numbers in increments of 3, 5, 10 or 20 in a low voice for another 8 minutes. This control version, therefore, had all of the same elements as the stress condition, without any social evaluative stress.

The next part to this study involved the participants who had completed the Trier Social Stress Test (both those in the stress condition group as well as those in the control group) playing a number of games against a partner drawn from the group of participants who did not do any version of the test. (The reason that these partners did not do the stress test as well was to avoid the 'misery loves company' effect.) Researchers wanted to know whether enduring social stress might lead to increased pro-social behaviour, and not only because the volunteers felt they had shared an experience with their interaction partners. The various games were designed to test how much the volunteers were willing to trust their interaction partner, how trustworthy they themselves were, and whether they would seek to share equally with their partner. Pro-social behaviour was indicated by higher levels of trust and more fair offers. In these games the volunteers and their partners could win real money.

Researchers first found that in the stressful condition participants' heart rates were higher and their levels of cortisol increased from the baseline measures, indicating that the social stress induction had worked. Most importantly, however, they also found that the volunteers in the social stress condition were more pro-social to their interaction partners. They trusted them more, were more trustworthy themselves, and were more likely to share equally.

Further evidence for this effect comes from a study in which researchers set up an elaborate field experiment allowing them to examine the impact of naturally occurring reminders of death.[12] They positioned a confederate who either talked audibly on a mobile phone about the value of helping, or about nothing in particular, in two different locations. At the first location the confederate stood in a cemetery, where passers-by could hear the conversation as they walked through the graveyard; while at the second he stood in the vicinity of the cemetery but where the cemetery itself was out of sight. As the passers-by continued walking they were presented with an opportunity to help a second confederate, who dropped a notebook while struggling with her backpack. Among those prompted to help, the number of participants who helped was 40 per cent greater at the cemetery than a block away.

The same researchers ran a second study in the same place, except this time they simply had a person sitting in a wheelchair who dropped a book in front of passers-by, again either inside the cemetery or a block away. The effect was the same. When in a context that reminded people of death, passers-by were more likely to help the person in the wheelchair (20 per cent more likely this time) compared to when they had no such reminders.

Overall the evidence shows that whether it is physical pain, social stress or background reminders of our own death, people naturally engage in more affiliative and pro-social behaviour thanks to these experiences.

BECOMING BETTER PEOPLE

So far we have looked at evidence which demonstrates that immediately before or immediately after an acute negative event people become more affiliative and more pro-social in their orientation towards others. These short-term effects are interesting and important; they suggest a variety of ways we might seek to build social connection with others. The findings suggest that it may be bad times rather than good which are most effective in achieving this goal.

One thing that is important to know is whether this pro-social response might also be apparent for those who experience ongoing and long-term adversity in their lives, rather than just acute and mild episodes of discomfort. Furthermore, whether this experience of adversity leads to longer-term effects, shaping people's characters rather than only their immediate responses.

This is the very question that Daniel Lim and David DeSteno from Northeastern University in Massachusetts sought to answer with two studies.[13] In the first, they asked volunteers to report on their past history of adversity, specifically whether they had experienced trauma associated with injury/illness, violence, bereavement, personal relationships, social or environmental stress, or from natural disasters – pretty much the whole gamut of possible adversities that one might experience. Next, they asked the volunteers to respond to questions designed to assess their levels of empathy, by asking them questions about whether they tended to feel sorry for other people who had problems in their lives. They also assessed the volunteers' levels of compassion, scoring statements such as 'It's important to take care of vulnerable people' or 'When I see someone hurt or in need, I feel a powerful urge to take care of them'. As an additional measure of compassion, they asked the volunteers, who had been paid $1 to participate in the study, if they would be willing to donate some of their earnings to the American Red Cross. The findings revealed that the more adversity people had experienced over their lifetime the more they felt empathy for others, and this in turn meant they were more likely to feel compassion for victims of misfortune and to donate their money to charity.

In a second study the researchers attempted to examine whether a lifetime history of adversity might indicate how people responded to others' needs within a more controlled laboratory environment. This time the volunteers were brought into the laboratory and were told they would be assisting the experimenters by observing another participant. In fact, the other 'participant' was a confederate and their behaviour was carefully modelled for the purpose of the study. The volunteer was told they would be observing the other participant's responses to a computer-based questionnaire live on their own computer screen. When the confederate entered they first completed a well-being questionnaire and indicated that they were feeling unwell at present, with the volunteer watching on their own screen as they responded. The volunteer watched as the confederate was then given the instructions that they would be randomly allocated to either a 15-minute task that was easy to complete or a 45-minute task that was very hard to complete. In fact, unbeknownst to the volunteer, the confederate was always allocated to the very hard task. Once the confederate 'learned' this they let out a soft, audible sigh and then called the experimenter to come over. They then engaged with the experimenter in the following conversation just outside the door so that the volunteer could overhear what was being said:

CONFEDERATE: I've been assigned to the red condition.
Hmmm . . . [pause] I'm sorry but is it possible for me to reschedule the experiment for tomorrow or some other time? I've not been feeling very well recently and I have a doctor's appointment at student health services in under an hour.
EXPERIMENTER [after some contemplation]: Unfortunately, we have limited research credit for this study. I'm not sure if I will have enough credits to reschedule you for a future session. But we can try to work things out and see if I can reschedule you for a future session. Nonetheless, it's entirely up to you on whether you want to stay or leave.
CONFEDERATE: Hmm . . . Alright then I'll stay to complete the experiment.
EXPERIMENTER: Thank you for helping us out. Please take your belongings and follow me to the next room to complete the experiment.

At this point the volunteer's computer prompted them to answer some questions about the assignment so far. Many of these were simply filler questions and unrelated to the study, but mixed in were some specific questions asking the volunteers about the level of sympathy and compassion they were feeling, on a scale from 1 (not at all) to 5 (very much). After completing this short survey, the volunteers were told the experiment had finished. However, they were also told that if they wanted, they could help the confederate finish their tasks in the other room.

Mirroring the findings of the first study, results indicated the more adversity a person had experienced during their lifetime, the more empathy they tended to feel towards others, and this in turn predicted their level of compassion for the confederate and the amount of time that they stayed back to help the confederate finish their tasks. What makes this study so important is that it links a person's history of adversity, drawing on a vast array of difficult experiences, to current small-scale, pro-social behaviour. It demonstrates that a lifetime of stress, pain and trauma can itself have long-term positive effects, focusing people on the needs of others.

Whether through pain, sadness or stress we tend to not only seek out support from other people, but we are also more likely to act in pro-social ways – tending to and befriending those around us. Is it no wonder, then, that the strongest forms of community are generally found in lower socio-economic areas – places where people deal with more of these types of experiences daily? It is here we find real community.

PAIN AS SOCIAL GLUE

This connection between pain and social support extends beyond the individual level into group contexts. Pain, fear, stress and even humiliation are central components of group initiation rites (think of the many hazing ceremonies conducted in American college fraternities over the years), and such negative experiences have long been incorporated into many rituals around the world, throughout history. This raises the question, what are these painful experiences achieving, and why have these kinds of rituals and group practices persisted over time?

A classic study by Harold Gerard and Grover Mathewson from the University of California in 1966 demonstrated how pain may enhance the value of group membership.[14] As part of the study, they invited college women to listen to a group discussion on sex. The catch was that to gain permission to hear the group discussion, the women were told either (a) they had to undergo a mild electric shock, or (b) they had to undergo a strong electric shock – and this time the experimenters really did electrocute their volunteers. In effect this modelled the kinds of practices involved in hazing and other initiation rituals, where people have to endure painful experiences before they are allowed to join a group. What they found provided new insights into why these rituals may have come to exist.

The volunteers who experienced the severe shock rated both the discussion and the other group members more positively compared to the volunteers who experienced the mild electric shock. However, this effect was evident only when the women were told that the shock was a necessary prerequisite to join the group discussion. When they were told it was unrelated to whether or not they could join the discussion, the severity of the shock no longer predicted their rating of the group.

To understand the findings of this study, we need to draw on a well-known psychological principle called 'cognitive dissonance'. This term refers to the feeling of discomfort that we experience when things don't make sense or don't fit together. A classic paradigm in which to demonstrate the effects of dissonance is to ask people to write an essay in support of something they personally disagree with. Imagine if you felt strongly that the death penalty should not exist, but had to write an essay in support of this type of punishment. What tends to happen is that, quite unbeknownst to those writing the essay, they tend to subtly shift their opinion to align with the essay they have just written. This is because we prefer our worlds to be consistent, and we have a strong motivation to resolve inconsistencies when they arise, even if that means shifting our opinions on things we care about. This same process explains the effects of the above study. When people endure some personal cost to join a group, such as undergoing electric shocks, their behaviour indicates that the group is of high value; we only endure personal costs for things we care about and which are important to us. In line with the motivation to resolve dissonance, our actions ensure we increase the value we place on the group. If we don't, then we are left with the unpleasant experience of acting in ways that are inconsistent with our attitudes.

This research casts a new light on hazing practices which might otherwise appear pointless. The unpleasant experiences in these settings increases the perceived value of joining the group. Now, these 'rituals' can go too far, leaving people traumatized and isolated. Yet hazing appears moderate compared to many older ceremonies. During the Hindu ritual of *kavadi*, performed at the festival of Thaipusam, participants engage in a range of painful acts such as inserting hooks through their skin, or knives or guns through gaping holes cut into their cheeks.

A recent study by the anthropologist Dimitris Xygalatas and his colleagues reported findings from their visit to the Thaipusam festival in Mauritius.[15] They surveyed both the participants of the *kavadi* as well as those who observed them (typically relatives). They also surveyed participants who undertook a less extreme ritual involving singing and collective prayer as part of the festival. At the end they asked participants a number of questions, including ratings for how painful the ritual activities were and how strongly attached to their Mauritian identity they felt. Finally, they offered participants the opportunity to donate some of the money they would receive for participating in the survey to the local temple.

The researchers found that participants who engaged in the painful *kavadi* ritual felt more attached to their Mauritian identity than those who prayed. They also found these same participants donated more of their money to the temple. Moreover, they observed that the participants' ratings of pain directly correlated with their donating behaviour: the more painful the experience, the more money they gave to the temple.

These effects were not limited to the participants themselves. Researchers also asked the observers of the ritual how painful they thought it would have been for the participants and gave them the opportunity to donate their money to the temple. They found the same pattern of results. The observers of the painful ritual gave more money to the temple than the observers of the less extreme ritual, and their ratings of how painful they thought the ritual would have been also correlated with their donating behaviour.

Xygalatas interpreted these findings as providing support to the long-standing ideas of the well-known sociologist Émile Durkheim, who argued that intense rituals promote co-operation. The Mauritian study showed – in a real-world context – that suffering (and some pretty extreme suffering at that) not only makes people more generous, but also increases identification with the community.

A further example of the ways in which painful rituals lead to group commitment comes from Australia. In Aboriginal and Torres Strait Islander cultures, it has been common for boys to participate in ceremonies that mark their initiation into manhood. These ceremonies often involve several painful practices, such as sitting close to fierce fires for prolonged periods, lack of sleep, or even the extraction of a tooth. An Australian anthropologist called Alfred William Howitt (1830–1908) directly participated in some of these ceremonies. Drawing from his own experiences he was adamant that their function was to increase group bonds. He stated that 'between the initiated there is . . . no reservation, but a feeling of confidence – I might even add almost of brotherhood'.[16] Such a painful and highly significant ceremony in a boy's life intensifies his bonds with other initiates, as well as all who were initiated before him, and this bonding stems from the shared nature of the painful ritual.

Painful initiation rituals are also seen in parts of Papua New Guinea. These rituals are so extreme that they have been termed 'Rites of Terror'[17] and involve burning and bodily mutilation. Such practices have been thought to enforce a political and religious community that is fixed for ever in the minds of participants. Those who go through this terrifying experience are believed to forge profound bonds of solidarity, and to develop a distinct identity that sets them apart as special or unique.

While the effects of painful rituals can lead to connection with others through the process of higher-level reflection (e.g. 'we all had the same powerful experience and now we are similar in important ways'), there is also evidence that these adverse experiences can lead people to align with each other at a biological level. Focusing this time on the ritual of fire-walking in the village of San Pedro Manrique in Spain, a group of researchers led by the Danish neuroscientist Ivana Konvalinka measured the heart rates of the fire walkers as well as those who were watching them.[18] What they found was that the collective ritual synchronized the heart rates of participants and observers, although they only found this effect for those who were closely related to each other, and it was not evident for unrelated spectators. We are more empathetic and engaged when it is a member of our own family in pain. More broadly, however, it also suggests that this empathy can lead us to share experiences, even at the biological level, in ways that harmonize us with those we love.

These same patterns, evident within age-old cultural rituals, are also apparent across a range of more commonplace contexts. For instance, the use of pain to enhance group bonding is often employed within the military. Soldiers typically undergo shared adverse experiences throughout training and combat, and such experiences foster camaraderie. The Canadian Airborne Regiment (CAR) is famous for putting their new recruits through hazing rituals, which may involve strong electric shocks, push-ups and induced vomiting. CAR members believe that such initiation activities prove the participants' readiness to join the group, as well as testing their commitment and loyalty. Those who do not participate are commonly ostracized. One officer describes the result of these shared experiences as 'a closeness unknown to all outsiders. Comrades [initiates] are closer than friends . . . closer than brothers. Their relationship is different from that of lovers. Their trust in and knowledge of each other is total.'[19]

Beyond the use of these practices in training, the trauma of war often works in the same way. This was observed in a longitudinal study of American war veterans.[20] Soldiers who had experienced combat, war trauma and the loss of significant others felt bonded together through these events. More recently, a field study of Libyan revolutionaries found that front-line fighters reported especially high levels of bonding between themselves and others in their battalion, compared to those who served in logistical support roles.[21] In fact, for half of the combatants surveyed, they felt *stronger* bonds with their fellow battalion members than with their own families. It is this same theme that is captured in the film *The Hurt Locker*. The protagonist is an American soldier who returns home to his family but cannot escape the feeling of being responsible for his co-combatants on the battlefield. Rather than staying to protect and care for his own family, he returns to fight with his 'brothers in arms'.

The effects of pain on bonding are not limited to cultural rituals, army training or wartime trauma. Enhanced co-operation and bonding are often evident in contexts where people experience large-scale disasters and emergencies. In 2010, I was working at the University of Queensland, in Brisbane, Australia. In December of that year Brisbane experienced its biggest flood in almost forty years, affecting the homes and businesses of over 200,000 people. Although we lived only minutes from the river, our house was on a small hill, so we were safe. We watched the floodwaters rise, covering many of the houses in our neighbourhood. Those who waited too long had to be evacuated by boat, seeing their homes and everything they owned covered in the muddy waters of the Brisbane River. Over the days that followed, we became acquainted with more people in our neighbourhood than ever before. We all had a common talking point. There had been common talking points before, such as when Queensland beat New South Wales in the rugby (a point of immense pride for Queenslanders), but nothing came close to bonding people in the way that this loss and misfortune had.

The aspect of this event that still conjures an emotional response, even as I write this, was the community reaction to the clean-up effort. Some 55,000 people volunteered over the weekend following the floods to help clean up the city. As a sign of the trust that had emerged within the community, people were letting strangers walk into their homes and help them to throw out their damaged furniture, or remove the layers of mud that had covered their floors. People were wandering the streets with brooms and mops looking for something they could do. I walked down to one of the flood-affected areas to help and was offered free water and food that people had brought to hand out to the volunteers. To be honest, it was easier to find a free sandwich and a cold drink than it was to find something to do – there were already so many people helping out. Ironically, one of the hardest jobs for the local authorities was not cleaning up, but trying to manage the sudden onslaught of manpower that had offered itself to assist.

A very similar effect was found in response to the 9/11 attacks on New York. A group of researchers analysed data from 605,454 people who volunteered between 7 August 1997 and 31 December 2001 through an online organization that matches volunteers with service organisations.[22] What they found was a significant spike in volunteering during the week of 11 September 2001 and extending to the two following weeks (see Figure 7). This indeed shows a sudden increase in pro-social behaviour in response to the devastation of 9/11, consistent with the same response I experienced after the Brisbane floods. Across America people not only volunteered more for organizations that were directly involved in responding to the crisis, but also for a range of other non-crisis-related charitable organizations. These included organizations focused on groups such as young people and the aged ('popular targets'), and issues such as education and hunger ('community activities'). This surge in volunteering did not stop there, as people also gave up more of their time to help those from marginalized groups such as gays, lesbians, bisexuals and even groups that would have been stigmatized in response to the attacks themselves, such as immigrants and refugee-support groups.

Pain motivates people to help each other, an effect that can thus spread and enrich communities. Whether it be observers of extreme rituals, or those who look on when others are affected by disaster and tragedy, these negative experiences trigger a desire to forget oneself and help those who have been affected. Moreover, this pro-social inclination becomes generalized; it spreads even to the non-affected, motivating people to become more accepting, more neighbourly and more caring.

These observations of community enrichment in response to disaster show that adversity can enhance group life; it can make groups function better. This made me wonder whether we could demonstrate

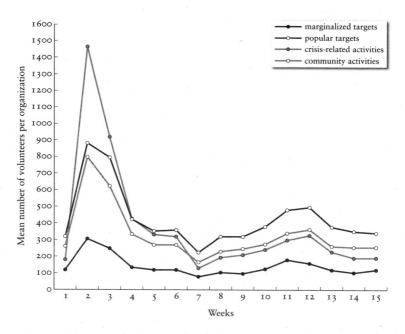

Figure 7. This graph shows the spike in the mean number of volunteers in the weeks after 9/11. Although this spike is most clear for activities related to the crisis itself, it was also evident for a range of other volunteering activities as well, including those involving groups that are traditionally 'unpopular' and the targets of prejudice.

this relationship between shared pain and group functioning in the lab.[23] To do this we had groups of students, who did not know each other previously, participate in the study. They were first introduced to the 'physical acuity' tasks. This time, in addition to the ice-bucket, participants also did another task. In the pain condition, they were asked to put their back against a wall and lower themselves until their legs were at a right angle. This position very quickly causes the thighs to burn, a sensation that builds the longer the position is held. In the pain-free condition, participants were asked to balance on one leg for 90 seconds – a challenging but non-painful task. After each group had performed these tasks in full view of each other, we gave them a brief questionnaire that, in addition to asking about the pain they had experienced, asked them a number of questions about how they felt towards the group members, including the extent to which they felt bonded with the other participants in their group, how much they felt they could trust them, and whether they felt a sense of loyalty and solidarity to the other participants. We found that the participants who had shared painful experiences together felt bonded by these experiences.

Encouraged by this finding, we decided to take things further. Having people indicate their attitudes on a questionnaire is one thing, but showing they will actually behave differently is the holy grail of psychological research. Therefore, rather than rating their feelings of bonding after the physical tasks, we now asked them to play a game in which they could earn money for themselves. The game involved choosing a number between one and seven across six different rounds. Choosing a seven meant they would earn $7.80, but only if every other group member also chose a seven. If someone chose a one, that person would receive $4.20 and anyone who chose a seven would only get $0.60 (see Figure 8). We gave payouts for each number, so choosing a higher number could earn volunteers the most money, but only if the whole group participated, while choosing a lower number protected your personal interests, but also undermined the interests of other group members. In essence, choosing higher numbers indicated a greater willingness to risk personal loss for the benefit of the group – a more co-operative choice.

Number chosen by you	Lowest number chosen in the group						
	1	2	3	4	5	6	7
1	$4.20						
2	$3.60	$4.80					
3	$3.00	$4.20	$5.40				
4	$2.40	$3.60	$4.80	$6.00			
5	$1.80	$3.00	$4.20	$5.40	$6.60		
6	$1.20	$2.40	$3.60	$4.80	$6.00	$7.20	
7	$0.60	$1.80	$3.00	$4.20	$5.40	$6.60	$7.80

Figure 8. Payout matrix for the co-operation game in experiments 2 and 3.

Those in the pain group chose higher numbers, on average 4.35, compared to those in the pain-free group, who chose on average 3.58. This was strong evidence that pain had actually increased co-operative behaviour between a group of strangers. Participants did not experience the pain *for* the benefit of the group or as a demonstration of group commitment (they did it as part of an experimental task focusing on their own physical abilities), yet they did share that experience and it bonded them.

One lingering question was whether our approach to inducing pain in the lab might have been confounded in some way. For instance, it could be that doing a challenging physical task together might have led to our effects. For this reason, we decided to change the task to something more representative of things we normally do. When I used to eat with friends at a Szechuan restaurant in Melbourne, we would order a plate of prawns buried in chillies, or the hot dishes with a side of cucumber, and a few cold beers to ease the burn. I always noticed that, as we progressed through the meal, the conversation would become more disjointed and the eating (and recovery after each mouthful) more frenetic. It was certainly a shared experience of pain. So, like all good experimental psychologists, we aimed to replicate this experience in the laboratory.

We asked participants to eat a bird's-eye chilli pepper, or as much of it as they could handle. Testing the strength of the various chilli peppers made several of our research meetings very entertaining. Our experience was not unlike that of the Danish National Chamber Orchestra who attempted to perform after each member ate the world's hottest chilli pepper (it's worth watching on YouTube). For a comparison condition, we asked another group to eat a butterscotch sweet. We told both groups that the task was for the purpose of a consumer-preferences study. After eating the chilli (or the boiled sweet) participants played the same co-operation game as before. Again, we found that after pain, participants chose higher numbers, an average of 4.33, compared to those who ate the sweet, whose average choice was 3.52 (see Figure 9).

Across these three studies, we provided experimental evidence that pain can be a powerful force that brings virtual strangers together into bonded and co-operative groups. This finding also echoed what I had observed in the real world, from my experience in the Brisbane floods to anecdotal evidence from war veterans or the survivors of atrocities; people become more caring and co-operative when they endure painful experiences. Not only is it true that what doesn't kill you makes you stronger; it also makes you better.

TENDING AND BEFRIENDING

There is abundant evidence showing that pain, stress and trauma – adversity in general – fosters community. This response has been exploited in a range of cultural rituals from *kavadi* to fraternity initiation ceremonies, and the persistence of these ceremonies and practices over time suggests that such effects are reliable – our painful experiences have bonded us together for centuries. Knowing these

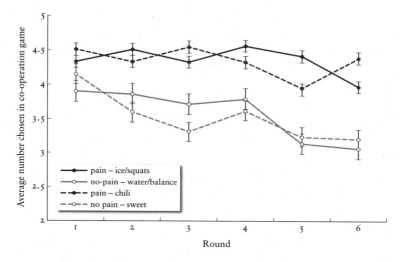

Figure 9. Mean choices for each round in the co-operation game in both co-operation experiments across all six rounds. Results are plotted separately for each condition.

effects of pain is useful, but it is also important to understand the psychological process that leads people to become more pro-social.

Shelley Taylor from the University of California has sought to understand this process. She began with the observation that almost every psychological theory would suggest that the predominant response to threat and stress is to fight or run away – to escape and protect ourselves; a response triggered by the amygdala (the threat-detection system in our brain). It is an evolved response that has kept humans, and most other animals, alive for millions of years. Yet, fighting or running away is inconsistent with helping, giving and affiliating. Taylor thus developed a theory that is widely known as the 'tend-and-befriend' hypothesis.[24] According to her theory, responding to conditions of stress by tending to offspring and affiliating with others ('befriending') is at least as common as our fight-or-flight response. Taylor's work was ground-breaking. Overturning the notion that it's 'every man for himself', this new theory argues that we respond to threats by helping one another.

In her work Taylor took the 'every man for himself' phrase literally, and predicted that tending and befriending is predominantly a female response. Taylor believed that this response would be most evident in females because they have a higher level of parental investment in their offspring. Accordingly, females would tend to their offspring and seek affiliation with social groups to reduce risk. She may have been partly wrong about this. If you recall the study earlier in this chapter led by Bernadette von Dawans, showing that social stress led to pro-social behaviour, you might remember that the team ran their experiment with only men. Von Dawans and her colleagues wanted to challenge Taylor's notion, and they did. Both women *and* men appear to have the tend-and-befriend inclination in response to stressful events.

In addition to fighting and running away, people also nurture their relationships with others, protect those they care about, and seek the security of their social groups when they experience threatening events. As a result, such events can lead to an improvement in group functioning and the strengthening of communities, outcomes that last well beyond the threats that produced them.

GETTING CREATIVE WITH PAIN

Having already observed that pain can bond people together and increase co-operation, my colleagues and I wanted to examine whether this effect of pain on groups could extend to strengthening groups in other ways. In particular, we wondered whether sharing a painful experience with others might generate the type of group climate that would allow for creative idea generation. There is a general consensus that in order for creativity to emerge within groups it is important people feel comfortable to express divergent and sometimes risky ideas. One way in which this process is frequently undermined is when people feel sensitive to potential negative reactions from others, such as being ridiculed or having others disagree rather than listen to ideas that are expressed.

We had a hunch that sharing a painful experience with other group members might lead to a more open, supportive and accepting group climate. This was for two reasons. First because, as detailed above, sharing a painful experience with others leads people to feel connected to and bonded with those others. Secondly, as I have also detailed above, seeing others in pain as well as experiencing pain oneself makes people more attentive to other people – because of our tendency to feel empathy for others as well as our need to affiliate with them. We reasoned that together these effects of shared pain would foster an environment characterized by reduced fear of negative evaluation and enhanced willingness to express divergent and novel ideas, and in this way would create a group climate that allowed creativity to emerge.[25]

To examine this, we had groups of between three and five people come into a large room. As in our other studies we initially asked them to do two tasks that either involved pain or did not. In the pain condition, the volunteers were first invited to take part in a 'consumer task' which involved eating as many bird's-eye chillies as they could (as before, these were intentionally very hot, and participants had water and yoghurt to cool their mouths). Following this, and after they had regained some semblance of composure, they were introduced to a 'physical task' where they were asked to maintain a leg squat (with their legs at a 90-degree angle) for as long as they could. In the control condition, the consumer tasks involved eating a butterscotch sweet and the physical task involved balancing on one leg – tasks that were similar but non-painful.

Next, each of the groups was given two different tasks to assess creativity. The first task is commonly used in studies on creativity and simply involves instructing the groups to come up with as many ideas for different ways to use a common house brick as possible. They are given two minutes to complete this task. Responses are then coded for creativity: how many different responses were generated, how many unique ideas were generated, and how many different categories of things are mentioned.

For the second task, the groups were told that they were to design a poster. We provided them with magazines they could use to cut out pictures or other images, and a range of stationery such as coloured marker pens, pencils, scissors and rulers, along with a piece of A3 paper as the canvas for the poster. The volunteers were told to 'create a work of art' with no further specifications about the types of art or how they should go about it. To determine the level of creativity demonstrated in each poster, we had a group of twenty university students rate each poster on nine points – how original, creative, imaginative, artistic, ingenious, innovative, unique, special and distinct each was – using a scale from 1 (not at all) to 7 (very much). Finally, to assess whether the volunteers thought their group had been creative, we simply asked them to rate the creativity of their group's responses on the brick task and how creative they felt their group's poster was.

We video-taped each of these sessions, which meant that at the end of the experiment we could have independent coders watch the recorded footage of the four tasks and rate how cohesive and cooperative each group appeared to be. They also rated each individual's behaviour, such as the extent of eye contact, talking, helping and encouragement each member of each group demonstrated, how comfortable each volunteer appeared to be, and how inhibited and guarded they seemed. All ratings were made on a scale ranging from 1 (not at all) to 5 (very much), and these ratings were then combined to give an overall measure of supportive interaction.

When we looked at the final ratings of how supportive the group climate was we found we had replicated the pattern of results that we observed in our earlier studies and provided additional support to the 'tend and befriend' hypothesis – when groups shared painful experiences (compared to when they shared non-painful but very similar experiences) they were more supportive of one another. Of course, this time the difference in supportiveness was assessed by independent judges rather than the participants themselves, providing more objective evidence that pain enhances pro-social interactions.

We also found that, to the extent that pain increased supportive interactions between group members, this in turn was associated with more creativity: a greater number of unique ideas for the uses of a brick, more creative posters as rated by the independent judges, and high levels of self-reported creativity in each group. Pain bonded people together and this in turn enhanced the creativeness of the groups' outputs. This was a new angle on the kinds of group-related processes that pain may promote: the ability to innovate. It certainly provided a new angle on the various ways in which teams responsible for innovation might seek to enhance their creative process.

While we have known for some time that enduring hardships can enhance performance – a fact routinely used in military training – the possibility that through creating a more supportive group environment such experiences can enhance other qualities, such as creativity, is unprecedented. This suggests that while luxurious corporate retreats might be good for morale, arduous corporate training exercises might be good for problem solving, creativity and a sense of unity. Our findings also fit with the belief that staying within our comfort zone is rarely a good strategy to foster creative outputs. We need to endure the challenge of sometimes stressful, novel and potentially threatening environments to foster true originality.

6

Finding Focus

The effort and concentration required, the tension and the stress resulting from the anguish, all increase the climber's awareness of both the immediate and more distant environment; everything is seen in a new light, with a clarity and the spiritual mobility, for example, that is also acquired through meditation.

Hans Selye[1]

The Sun Dance was one of the most spectacular and important cultural ceremonies of the North American Plains Indians in the nineteenth century. It was outlawed in 1904 by the United States government, in part because settlers at the time found the practices involved in the Sun Dance ceremony to be overly gruesome (although the ban has since been rescinded). Rarely spoken about, the ceremony represents the continuity between life and death, allowing for the spiritual rebirth of its participants and their relatives as well as the regeneration of the living Earth. It can last for between three days and a week and is typically agonizing. Self-inflicted torture, endured as part of the ceremony, represents death, and it is only through this process that a person can be symbolically resurrected. The dancers are believed to be reborn, mentally, spiritually and physically, through their suffering.

Sun dancers typically fast for the entire period of the ceremony and are fastened to a pole by ropes that are connected to hooks, eagle claws or rawhide thongs pierced through their chest. The dance involves moving forward and backward from the pole, slowly placing more and more pressure on the piercings. Many dancers collapse, and when they do they are believed to be given a vision for the guidance and renewal of their tribes. The dance culminates with the participants running backwards away from the pole so the hooks rip through the flesh on their chest and set them free of their ropes. Manny Twofeathers, a native Indian elder, describes his personal experience of the Sun Dance in his book *Road to the Sundance*.[2] He recounts the sharp and intense pain of having the piercing bones inserted into his chest and then a feeling that he had 'lost all sense of time'. He recounts how the agony led him to feel closer to the Creator and how he felt like crying for all the people who needed his prayers, those who were hungry or sick. He describes the process of leaning back on his ropes and the intense hurt, the raw ache that reached all the way down to his toes, and how when he looked down at his piercing bones he saw the faces of his children. In describing this experience he recounts: 'It felt glorious and explosive. The energy was high and brilliant.' Finally, Manny Twofeathers describes the process of running back from the tree to which he was tied, faster and faster until he hit the end of the line and heard his flesh 'tear, rip, and pop'. He describes this as a moment of elation.

The Sun Dance serves as a traditional pathway to renewal and spiritual enlightenment. It is through the experience of suffering that sun dancers feel they are able to transcend the physical world and receive visions or spiritual insights. This use of intense suffering to achieve spiritual enlightenment is not limited to this one ritual, however. Vision quests, also practised within Native American cultures (for example by the Cree Indians), involve young men being sent out to wander for days through the wilderness. During this time, they fast, avoid sleep and concentrate on their quest until their minds become exhausted, leading to hallucinations and visions. There are many other examples, such as the Good Friday ceremony in the Philippines, where volunteers re-enact the crucifixion of Christ and allow themselves to be nailed to a cross.

Experiences of spiritual transcendence arising from pain have been directly observed by researchers studying the *kavadi* rituals in Mauritius (referred to earlier). The participants commonly express symptoms of dissociation, such as amnesia, absorption and depersonalization.[3] A similar account is evident in the writings of St Teresa of Ávila (St Teresa of Jesus), a sixteenth-century Spanish nun who suffered greatly from illness. She describes it was through her suffering she could achieve a higher level of union with God. Indeed, she began inflicting mortifications of the flesh upon herself to heighten her spiritual experiences. Such practices commonly range from wearing shirts made from coarse animal hair, sometimes with wire or twigs for added discomfort, to self-flagellation with whips. For St Teresa, this was experienced as an ecstatic state, where the feeling of being in the body disappears, sensory activity ceases, and memory and imagination are entirely absorbed in God.[4]

This chapter will explore why pain has been exploited within these cultural and religious rituals for so long, and what it is about the experience of pain that allows for a sense of spirituality and enlightenment. Through this, it will become apparent how pain may give rise to transcendent experiences people commonly seek out today in Western societies via practices such as meditation or mindfulness.

THE BENEFITS OF TUNNELLING

In a recent book entitled *Scarcity*, Sendhil Mullainathan, an economist from Harvard University, and Eldar Shafir, a psychologist from Princeton University, detail what they refer to as the 'scarcity mind-set'. When people experience resource pressures, whether they're short of money, time, friends or food, they tend to automatically focus their attention on using what resources they already have available most effectively. While this focus may increase productivity and efficiency, it also comes at the cost of what they refer to as 'tunnelling'. That is, the tendency for people in a scarcity mind-set to miss out on important peripheral details in their decision-making. In an effort to only give full consideration to that which is in the tunnel – and therefore to what appears to be the most pressing concern – they no longer attend to other, broader factors. Understood in this way, scarcity serves to narrow attention and focuses us on what matters most.

Pain works in a very similar way to scarcity. As an evolutionary alarm signal, pain warns us of threat and danger, and it serves this function ruthlessly; attending to pain takes priority over almost all other goal-directed behaviour. Imagine making a sandwich in the morning as you are rushing to get your children ready for school and the knife slips and cuts your finger. Right then and there everything stops. There are many other competing demands for your attention in this moment – namely children and time. Still, your finger becomes the principal concern. It is in this way pain narrows our attention; specifically, it narrows it to our immediate sensation, and the mind becomes completely occupied in dealing with its causes and consequences.

What this demonstrates is that when pain is present, we are focused on it, and mostly to the exclusion of all else: it literally captures our attention. Several studies have now shown that patients who suffer from chronic pain are worse at performing tasks that require high levels of attention, the reason being their symptoms are constantly redirecting their attention away from the task at hand.

In one such study,[5] chronic pain patients were required to complete a task: they were asked to name the largest digit (6 in the example in Figure 10) or the largest number of digits (9 in the example). The researchers grouped the volunteers according to their self-reported level of pain so there were two groups – high pain and low

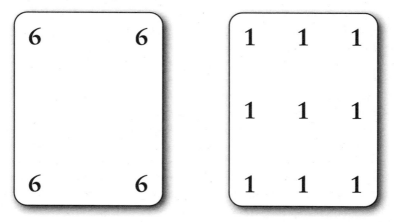

Figure 10. Two examples of the materials used in the study, which were presented in pairs. Naming the largest digit is easier – it requires fewer cognitive resources – than naming the largest number of digits, which requires the participant to concentrate and count.

pain. They also included a control group of volunteers against which to compare these patients. What they found was that the patients who self-rated their pain as high were significantly slower to determine the largest number of digits compared to the low-pain group and the control group. There were no differences when simply naming the largest digit, because this required fewer cognitive resources and therefore less attention.

This same effect has been found when volunteers are exposed to a single painful stimulus within a laboratory context; they too were worse at performing cognitive tasks.[6] So whether it is chronic long-term pain or acute temporary pain, these experiences draw on our cognitive resources, narrowing our focus to what is happening to our bodies and limiting our capacity to focus attention elsewhere.

The narrowing effect has two benefits. The first is related to survival. Both scarcity and pain symbolize a type of threat that triggers the fight–flight response. We focus almost exclusively on this threat as we need to allocate significant cognitive resources to ensure we escape danger. These high arousal and negative emotional states are also opposite to those which psychologist Barbara Fredrickson argues facilitate a 'broaden and build' mind-set.[7] Experiencing low arousal and positive emotional states gives us licence to sit back and take a broad view of the world. But threats cause us to narrow our field of attention to the immediate context – what is happening right here and now. We have developed this tendency because over time it has proved effective for survival; we can respond better to threats when we are acutely aware of them.

It is through this narrowing of attention that the second benefit arises. When our minds are focused on just one thing we experience a different state of awareness, one that is often sought out through practices such as meditation or mindfulness. The aim of these practices is to quieten the mind, to limit its tendency to flit about from one worry to the next. By forcing our attention to focus on just one thing, threats provide a distraction from other concerns. Of course, as reviewed in Chapter 3, research on benign masochism shows how negative events can signal threat even within safe contexts, and this conflict between threat and safety facilitates enjoyment. This also illustrates that we need not be faced with real danger for our bodies to respond to the threat signals that come from a variety of negative experiences. Even within safe contexts, these signals capture our attention (just think of how hard it is to count backwards from one hundred in sevens when you are immersed in freezing-cold water) and serve to clear our minds of more worrying but less immediate concerns.

If you consider the various practices through which people commonly seek out meditative or mindful states, they all tend to have this same quality: a sole focus. Practitioners teach people to quieten their thoughts by turning their attention to an immediate sensory experience. This may be achieved through focusing on a single thing – such as an object, thought or sensation – or by maintaining an ongoing awareness of the present moment.

Beyond training the mind, this same focus can be achieved through training the body. Within yoga or martial arts, physical movement is employed to focus attention on the body and often in ways that involve discomfort. In yoga, the Virisana, or 'hero pose', can be very painful (even for experienced practitioners). It involves sitting on the back of one's heels while keeping one's toes tucked under. It feels awkward and unnatural. As one experienced yoga practitioner describes, 'physical challenge, combined with mental focus and regulated breathing, can do much more than build strength and flexibility. It can be a way of opening up the field of body/mind . . . it is transformative.'[8]

This effect of physical discomfort or threat is elegantly captured in a quote from an S&M practitioner, who states that 'A whip is a great way to get someone to be here now. They can't look away from it, and they can't think of anything else.'[9] It becomes all about the immediate sexual experiences, and in this way pain achieves what many sex therapists try to get their patients to do – to use a 'sensate focus'.[10] When your mind wanders during sex, or when you start worrying about what your partner thinks about your love-handles (or, worse still, what you must do for work tomorrow), these distractions reduce arousal and inhibit sexual performance. Re-focusing the mind on immediate bodily sensations (as opposed to objective self-awareness or concerns about the future) can re-awaken sexual arousal, just as pain used in the practice of S&M may increase it.

Elaine Scarry also noticed this, albeit in a very different context.[11] Scarry provides an erudite and in-depth analysis of torture, and argues that torture works by exploiting this same effect; it distracts us from other elements of our experience, but also literally makes us feel as though ourselves and the world around us have disintegrated:

> It is the intense pain that destroys a person's self and world, a destruction experienced spatially as either the contraction of the universe down to the immediate vicinity of the body or as the body swelling to fill the entire universe. Intense pain is also language-destroying: as the content of one's world disintegrates, so the content of one's language disintegrates; as the self disintegrates, so that which would express and project the self is robbed of its source and its subject. World, self, and voice are lost, or nearly lost, through the intense pain of torture.

Scarry finds that pain can obliterate the contents of consciousness, and annihilate the objects of complex thought, emotion and perception. All the while pain itself maintains an almost complete resistance to objectification – we struggle to even put it into words and communicate it so others can understand. It is precisely this ability to experience the world without evaluation or objectification, to remove awareness of complex thought or the perception of other things, and to transcend our own self-reflection in the service of narrowing our focus to one single thought, object or experience that comes into play when trying to achieve mindfulness. While torture may achieve this to the point of annihilation, lesser and more controllable discomforts can promote this kind of experience in more moderate and beneficial ways.

People commonly seek out meditative or mindful states through quiet reflection, yet threat-signalling experiences provide an alternate, albeit more brutal, route. Like a short-cut to mindfulness, they instantly draw us into a singular focus. It is due to the capacity of such experiences to command attention, to remove awareness of complex thought, and to reduce higher-order reflection on the self that they are often experienced as providing a form of enlightenment or spiritual transcendence. If you consider the various practices that are thought to provide spiritual enlightenment, they all have these qualities in common. Beyond the various rituals detailed above, people also engage in meditative prayer, chanting, singing or dancing to achieve a sense of spiritual transcendence. These acts, through repetition, fix one's mind on a singular thing, and limit the potential for distraction.

HOMEOSTASIS AND AWARENESS — RIGHT NOW!

If threat signals can provide an effective avenue for attaining states of mindfulness, then they should not only provide a singular focus but also ensure that this focus is on the present. Practitioners emphasize this element of mindfulness training, directing students to focus on the present, to be aware of each passing moment. The reason for this is that the worries and concerns which cloud our minds are rarely within the immediate environment; they tend to be about relationships or disputes, work or financial concerns. When we worry, it tends to be about the implications of things that have already taken place, or the possible outcomes of future events. Neuroscientific evidence shows that pain and other forms of threat can also achieve this focus on the present. Beyond fostering a singular focus on the causes and consequences of the threat, such experiences also ensure our minds are fixed on the present moment, on what is happening right now.

In a highly cited review paper, leading neuroscientist Bud Craig explored the neural underpinnings of pain.[12] He argues that the insula, a part of the brain involved in the processing of painful experiences, is also the part of the brain responsible for our awareness. Specifically, he suggests pain and other interoceptive experiences (interoception refers to the extent to which we are aware of the current state of our body) activate the anterior insula cortex, and it is the activation of this neural region that facilitates an awareness of what is happening right now. In fact, Craig suggests it is the activation of this region that gives rise to awareness in general. He claims that without interoceptive experiences activating the anterior insula cortex we would not be conscious of anything at all.

Craig's theory of human awareness is not limited to physical pain, but extends to a range of experiences, all of which are associated with the physiological state of the body, and all of which trigger interoceptive awareness. Craig views pain as a homeostatic emotion,[13] meaning that like hunger, thirst and temperature (feeling too hot or cold), it sends a message to the brain about the body's current homeostatic status. Homeostasis refers to the state of the body when it is in balance and functioning optimally – when it not too cold or too hot, and has sufficient nutrients and liquid. When these basic physical needs are not met, the body enters a state of homeostatic imbalance. According to Craig, it is only when our body is in this state of imbalance (however minimal) that we become aware of our physical selves.

According to this theory, people should be more connected to the present when their bodies are in a state of homeostatic imbalance. It is no wonder, then, that within religious rituals people not only exploit pain, but also other avenues through which they can create such an effect. Fasting is a good example. This is common to a range of religions, including the Christian festival of Lent and the Muslim month of Ramadan. Ascetics are another good example. These individuals, whether they are from the Hindu, Buddhist or Muslim faiths, engage in the practice of self-deprivation. The word is derived from the ancient Greek *áskēsis*, meaning 'exercise' or 'training', and ascetics believe indulging in sensual pleasures blunts spiritual awareness, and enlightenment is only achieved through self-deprivation. This intuition is consistent with Craig's theory, which would concur that it is homeostatic *imbalance* or discomfort, as opposed to homeostatic *balance* or comfort, which serves to increase consciousness of what is happening, right here and right now. It is by exploiting this effect that people experience feelings of spiritual transcendence or enlightenment.

To return to the Sun Dance festival, the American Plains Indians traditionally incorporate a sweat lodge into their experience, where participants are exposed to high levels of heat, steam and dehydration. Creating this experience is something that needs to be undertaken with a certain level of caution. While eliciting signals of threat can be achieved in relatively safe ways, they can easily be taken too far. Sadly, the fact is some *have* taken these practices too far. In 2009 a self-help guru, James Arthur Ray, had people pay $9,000 for a 'spiritual warrior experience'. A highlight was the sweat lodge, which he told participants would make them feel like they were going to die, like their skin was going to fall off, but afterwards they would emerge as a new person. Unfortunately, he pushed his students too far, leading to the deaths of three people and the injury of others.

TRANSCENDING THE EGO

To more fully understand how putting the body into a state of imbalance can lead to an altered state of consciousness, and in turn to a
sense of enlightenment or spiritual transcendence, we should consider other ways in which people seek out meditative states. One
common approach is the use of floatation tanks in which there are no
sounds, no light, and the water temperature is carefully controlled to
ensure the body is neither too hot nor too cold. A floatation tank, in
effect, seeks to ensure there are no homeostatic imbalances and therefore nothing by which to become distracted. Yet, rather than
becoming absorbed by concrete worries and concerns, people often
report extremely vivid mental images and visions.

People experience these different states of consciousness because
without any sensory input one's experience of the self becomes distorted; there is no longer any sensation on which to focus or assess the
physiological condition of the body and one's mind is left to wander
virtually 'unembodied'. As Craig suggests, without any interoceptive
experiences activating the anterior insula cortex we may cease to be
consciously aware of ourselves at all. All of this results in a kind of
dissociative state where people are no longer focused on their self-
oriented worries and concerns.

This contrasting pathway towards achieving mindfulness provides
for interesting reflections on the ways in which homeostatic imbalance
may achieve these same ends. Without sensory input our sense of
the phenomenal-self (ourself as existing in the immediate moment)
becomes profoundly distorted. When our bodies are thirsty, hungry or
in pain, they achieve the same experience, albeit in a different way;
these imbalanced states drown out other self-relevant thought and in
this way also produce a profoundly different experience of the self. Just
as a lack of interoceptive input leads us to abandon our ego, so do
impending bodily threats. For those who teach meditation and other
pathways towards spiritual transcendence, it is this loss of self-
awareness or ego which is often viewed as a necessary element of the
process. As Swami Vivekananda, an Indian Hindu monk, said: 'Unless
one renounces the ego, one does not receive the grace of God.'[14]

FINDING FOCUS BY FINDING OUR LIMITS

People seek mindfulness through meditation because this state of awareness is both healthy and pleasant. It is a great way to counteract stress, and even to reduce longer-term difficulties like anxiety and depression. As detailed above, there are many rituals, some of which can be quite extreme, that achieve this same effect and leave people with a sense of spiritual transcendence. Yet there are also many common and less ritualized contexts in which states of mindfulness can be achieved through discomfort.

One domain in which similar effects have been observed is in the pursuit of extreme sports. This is captured in the quote below from the mountain climber Lionel Terray, reflecting on the moment he found himself in difficulty: 'My personality left me, the links with the earth were severed; I was no longer frightened or tired; I felt as though transported through the air, I was invisible, nothing could stop me, I'd reached that state of intoxication, of dematerialization . . .'[15]

This link between states of mindful awareness and the sense of 'being on the edge' in extreme sports helps us to understand why people seek out and enjoy these experiences. In her phenomenological investigation, Carla Willig finds that suffering, mastery and skill, and being in the present, are all factors that motivate people to take part in extreme sports.[16] People become aware of themselves in a new way, in a way that transcends their reflection on their future or past selves, and which brings them into direct contact with their moment-by-moment experience of the world.

This idea is reflected in a comment by Keira Henninger, an ultra-marathon runner. In 2013 both Keira and I were interviewed on *The Huffington Post Live* about the upside of pain. Keira detailed how she spends most of her weekends running for hours on end. When asked why she enjoyed running such long distances, she responded: 'Anybody can go out for a run and it can clear your mind, take away the stress of life, but running 100 miles – it clears your soul, it's the best thing in the world.'

Each year the Sri Chinmoy Self-Transcendence 3,100 Mile Race takes place in the borough of Queens, New York. This is an ultra-marathon which has a time limit of fifty-two days and is run each day between 6 a.m. and midnight. It is named after Sri Chinmoy, an Indian spiritual teacher who taught that physical pursuits such as running or weightlifting are key pathways through which people can achieve spiritual transcendence. Pounding the pavement for fifty-two days not only demands a strong mind and plenty of focus, but the activity itself becomes meditative – mile after mile, breath after breath, one foot after the other. Repetition has a powerful effect.

GLOBAL EMOTIONAL MOMENTS

Threatening or uncomfortable experiences can make time appear to stand still, providing another avenue through which such events can generate alternate states of awareness. This effect on time perception is also captured in Craig's model. Specifically, he refers to 'global emotional moments', momentary states of consciousness of our entire body. He argues that these 'moments' build up across short bursts of time so we are able to maintain a sense of ourselves existing within a particular situation. The continuity of experience is maintained in a kind of global awareness of ourselves prior to encoding in memory.

Craig suggests emotionally intense experiences take on a disproportionate value and clarity in the mind because the brain registers bodily states more frequently when potential threats exist in the environment. This is similar to how we think about 'flash-bulb memories', which are highly detailed, exceptionally vivid 'snapshots' of the time when an emotionally arousing event occurred. For instance, many people can recall exactly what they were doing when they first heard of the shootings in Norway by Anders Breivik or when Princess Diana died. In this case, however, vividness is not an aspect of memory but is evident in how we are currently experiencing the world – and time appears to stand still. It explains why spending 30 seconds submerged in ice-water feels a lot longer than 30 seconds lying on a hammock in the sun. The emotional intensity of the ice-water event means that we build up a richer representation of the experience, and this leads to the subjective perception that the experience lasts for longer than it actually does.

STRENGTHENING MENTAL CONTROL

Meditation requires mental control. It is in large part about learning to focus one's attention where it is wanted, and this can be tricky – our minds tend to slip off into unwanted thoughts in a flash. Meditation trains our ability to manage our minds, providing us with a powerful tool that helps us to better regulate our emotions. Importantly, enduring uncomfortable experiences can also strengthen this capacity.

If we return to the quote from our yoga practitioner above (page 131), it highlights that by maintaining a painful yoga posture we draw on our ability for 'mental focus'. While trying to maintain our attention on breathing in meditation can be tough, so can trying to maintain our composure during an experience of pain. To achieve this, we need to inhibit the very message pain is sending to our brains – which is to escape the situation. Pain is like a guy with a megaphone yelling at us 'get out, get out now', but to endure pain we need to stay calm and inhibit our desire to respond in exactly this way.

The human ability to inhibit unwanted thoughts, feelings and behaviours is an important skill that allows people to act in accordance with their goals. Given that the inhibition of these capacities needs to occur together and simultaneously to be effective (it is not much use inhibiting your thoughts if you still behave in the wrong way, or react emotionally), it would make sense that control over all these impulses would be regulated by the same neural mechanisms. What this also means is that intentionally inhibiting impulses in one domain may have a 'spillover' effect on other domains. Indeed, recent research has begun to show just this type of effect, and it turns out the same brain region (the right inferior frontal cortex) is involved in inhibition across a range of different processes involving cognition, emotion and behaviour.[17]

An especially intriguing example of this spillover comes from a series of studies investigating the effects of bladder control on impulse inhibition more broadly.[18] In the first study the researchers had 193 volunteers complete a Stroop task. This involves either stating a word (the name of a colour, e.g. orange) as it is written, which is the dominant response and does not require inhibition, or saying the name of the colour in which the word is written (e.g. black), which requires the inhibition of the dominant response. It is much easier to read the words than name the colours. Researchers also asked the volunteers to indicate the urgency with which they currently needed to urinate, from 1 (not urgently at all) to 7 (very urgently). What they found was that urination urgency was related to better performance on the colour-naming element of the Stroop task. The more the volunteers needed to inhibit their urge to urinate the better they were at inhibiting their tendency to read the words rather than say the name of the colours.

In a second study, the researchers had 102 university students either drink five cups of water (the high bladder-pressure condition) or just sip small amounts of water (the low bladder-pressure condition). Approximately 45 minutes later, the volunteers were asked to choose between receiving €16 tomorrow or €30 in 35 days' time. Those who had drunk more water and needed to urinate more urgently were more likely to choose the delayed gratification option. This demonstrated that the inhibition of the need to urinate had spilled over to the inhibition of the desire for a more immediate but smaller reward over a longer-term but larger reward.

Encouraged by these results, another group of researchers sought to extend these findings.[19] In this study the researchers again asked half their volunteers to drink a large amount of water, and the other half a small amount. After a break, they asked the volunteers to engage in an interview with another person where they were either instructed to tell the truth or to lie. These interviews were recorded and then rated by a third person, assessing the presence of behavioural cues and forming their impression of whether they thought the volunteers were telling the truth or not. The researchers found it was the volunteers who had consumed a large amount of water, and needed to inhibit their desire to urinate, who were also the best liars. The people rating the interviews more often thought the liars who needed to urinate were telling the truth compared to the liars who did not need to urinate.

It is important to note the apparent inconsistency between the effects of inhibiting the urge to urinate on enhancing task performance here, compared to the effects of chronic and acute pain on diminishing task performance as reviewed above. On the one hand, discomfort should narrow our attention and interrupt our ability to focus on other tasks. On the other, intentionally inhibiting the desire to extinguish an uncomfortable experience appears to enhance performance. While these two effects of discomfort appear to run in different directions, they also highlight two avenues through which we can enhance the focus of our attention. The very nature of homeostatic imbalance (whether caused by pain, thirst, hunger or the need to urinate) provides a powerful source of distraction, which can be exploited within a variety of contexts. Yet, added to this, when we intentionally inhibit the urge to escape from such experiences, this effortful inhibition 'spills over', allowing us to better self-regulate within other domains. As with the example of yoga above, it is through intentionally enduring uncomfortable physical postures (and inhibiting the desire to remove that discomfort) that people can train the mind to focus where they want it – on the act of meditation. While the available research suggests two effects that appear to be in juxtaposition to one another, it also provides two explanations for how discomfort may keep our minds focused.

CHANGING PAIN INTO PLEASURE

Based on the foregoing analysis there is solid evidence that pain may provide a kind of short-cut to mindfulness. Threatening or uncomfortable experiences can distract us from other worries and concerns that cloud our minds. Of course, strictly speaking our focus is still on something unpleasant. As reviewed above, a dip in cold water, fasting, or the feeling of fear when rock climbing may still bring some enjoyment, but it would be even better if through enduring these experiences we could increase our capacity to engage with pleasure. Happily, there is good reason to expect this is also a likely outcome.

Here, again, the idea of pain-offset is important – the moment at which the pain stops. As reviewed earlier, pain has the capacity to increase the experience of pleasure when it stops via the experience of relief. In part, this is because our brains continue to release opioids, which not only dull pain but also enhance pleasure. A similar process should also occur when it comes to the effect of pain on narrowing our focus of attention. When pain stops, especially if it stops suddenly, we are still aware of our sensory experiences in a way we were not before. It is pain that brings us into the moment, but once it has ceased we are left there for a period, more in touch with our sensory experience of the world than we were before.

It was this possibility that motivated a series of studies my colleagues and I conducted during my time at the University of Queensland. We reasoned that if pain really does put us in touch with our immediate sensory awareness – in a meditative or mindful way – it might enhance other kinds of sensory experiences once the pain has ceased. That is, could the sudden boost in immediate sensory awareness caused by pain carry on, meaning our next sensory experiences might be more intense?

We tested this idea by running a series of studies examining people's experience of taste.[20] The first study we ran focused on the most critical question – do we enjoy pleasant sensory experiences more after pain? A good example of the phenomenon might be the enjoyment of a thirst-quenching cold drink after a tough basketball match, or a mug of hot tea after coming in from the cold. Of course, it is also important to consider that this increased sensory awareness would be unlikely to persist for a long time. In fact, it is quite likely that within minutes our minds may begin to wander on to other thoughts about the past or the future, and our immediate sensory experiences may be less apparent to us again. To test this possibility we had volunteers either write an essay for 10 minutes about something they did yesterday and then had them place their hand into a bucket of ice-water for as long as they could (recent pain group), or they did those same two tasks in the reverse order – the ice-bucket then the essay (delay control group). After this, volunteers were asked to eat a chocolate biscuit and rate how much they enjoyed it.

We discovered the biscuit was more enjoyable for those who ate it immediately after pain, compared to those who ate it immediately after writing the essay but who experienced pain 10 minutes earlier. On a scale from 1 (not at all) to 7 (very much so) the 'recent pain' group gave an average rating of 6.03, compared to the 'delay control' group who gave an average rating of 4.97. Eating the chocolate biscuit immediately after a painful experience increased people's enjoyment by a whole point compared to eating it after a 10-minute delay.

Now it's natural to assume that contrast – this idea of relativity discussed earlier – is what explains the levels of enjoyment. That is, any positive experience would be more enjoyable by contrast to the negative experience of pain. We also realized we needed to go beyond biscuits to really nail this effect. So, in our next study we aimed to demonstrate it was not just pleasant sensory experiences but also unpleasant ones that were felt more intensely after pain. If it was just the contrast with pain that was driving our findings, then we would expect pleasant experiences might be more pleasant, but unpleasant experiences would be less unpleasant. Imagine failing an exam just after receiving news that your grandfather had passed away. The failure of the exam would seem a whole lot less significant, and probably more manageable, but this would not provide evidence for increased awareness, just the effect of relative contrast.

In our second study, our participants were given the ice-bucket task and then asked to drink four cups of liquid and rate how intense the flavour was. The four cups contained four different tastes – sweet (water mixed with sugar), salty (water mixed with salt), sour (water mixed with lemon juice), and bitter (tonic water). Those in the pain group rated all four tastes as more intense – on average 4.46, compared to 3.71 in the no pain, control group. In fact, we found the largest difference in flavour intensity was the salty liquid (perhaps the most unpleasant-tasting solution), so simple contrast effects could not explain our findings.

Encouraged by these results – and equally inspired by an episode of

Heston Blumenthal's cooking show where he tested taste sensitivity among aircraft passengers – we designed a third study. This time we wanted to know whether people were actually better at detecting flavours after experiencing pain. Using a series of small cups in three rows of ten, we carefully inserted very specific amounts of industrial flavouring into each cup, with no flavouring in the first cup and the most intense flavouring in the tenth. After the ice-bucket test, participants tasted each cup in order until they guessed the correct flavour. Those in the pain group were quicker to guess the correct flavour, on average correctly identifying the flavour on the eighth cup, while people in the control group on average only guessed the flavour by the tenth cup.

Overall our findings demonstrate that pain can increase our enjoyment of a subsequent sensory experience, and that this occurs because we are more sensitive and responsive to these experiences after pain. Our studies provided evidence that, just as mindful meditation may increase a person's engagement with, and enjoyment of, their sensory experiences, so can pain.

WHY CHOOSE PAIN?

Together the evidence suggests that pain can capture attention and increase awareness of our immediate sensory experience – it connects us to the world around us. Furthermore, our increased awareness is not limited to the experience of pain. When it ceases, this awareness can make other experiences more intense and more enjoyable. Still, surely there are better ways to achieve this! Why would I choose the shock of a cold shower, the exhaustion of a hard workout, or the fear felt when speaking in public over other practices (such as mindfulness) that can also increase awareness?

One reason to choose these uncomfortable experiences is that something like meditating is difficult and takes time and practice. Our minds tend to be like badly behaved children racing from one thought to another. This is why many people spend their time and money on classes or books teaching them meditation or mindfulness techniques – yet even with this help, achieving voluntary control over our attention can be fraught with failure. If you have ever tried meditating you will know it can be a very frustrating experience. Your mind wanders off just when you don't want it to, and then you fall into the trap of thinking about how bad you are at meditation instead of actually meditating, a process triggering that frustration loop all over again. It is for this very reason that one of the key skills taught in meditation classes is how to deal with feelings of frustration when they arise. Our minds are natural wanderers, and we find it hard to stop them.

Providing some insight into how and why our minds tend to wander is a series of findings over the past fifteen years on what happens when the brain is in a resting state. The first was reported in a paper published in the *Proceedings of the National Academy of Sciences* in 2001.[21] Marcus Raichle and colleagues found that there is a distinct pattern of activation in the brain when there are no goal-relevant mental tasks that need to be completed, and when there are no stimulus inputs such as sights or sounds. Building on this work, in 2007 Malia Mason and colleagues published a paper in *Science* showing this default network was related to mind-wandering, or what they called stimulus-independent thought (thinking about something that has nothing to do with what we can see, hear, touch, taste, smell or feel).[22] Findings revealed that when our mind is 'resting' it is actually not resting on the present moment. Rather, it is wandering off onto a multitude of thoughts, memories or any other type of mental content that is unrelated to what is currently happening in the immediate environment. The researchers found mind-wandering was most evident when people were engaging in tasks they had practised habitually, such as washing the dishes or commuting, and which therefore required little mental attention. While the reason for mind-wandering is not clear, Mason and her colleagues suggested our minds may wander simply because they can. Our mental capacity to represent abstract concepts and to time-travel into the past and the future is an important ability that sometimes distracts us from the here and the now.

Although mind-wandering may sound like a pleasant pastime, this process tends to compromise our happiness. In another paper published in *Science*,[23] Matthew Killingsworth and Daniel Gilbert investigated whether mind-wandering is related to more or less happiness. To do this they developed an iPhone app that allowed them to contact participants through their phones at random moments throughout the day. They asked more than 2,000 participants three key questions: 'How happy are you feeling right now?'; 'What are you doing right now?'; and 'Are you thinking about something other than what you are currently doing?'

What they found was that people's minds wander frequently. In 46.9 per cent of each of the moments sampled, the volunteers reported their minds had been wandering. In fact, it did not appear to matter what people were doing at the time; across all the different activities that people reported being engaged with, their minds still wandered 30 per cent of the time. The only time mind-wandering occurred very rarely was during sex. The sex therapists would be happy – and especially as the researchers also found people were *less* happy when their minds wandered, and this was the case no matter which activities the person was engaged with at the time. It did not seem to matter what people were doing, they were less happy if their minds were not focused on the task at hand, even when that task was itself not especially enjoyable. As the authors note, 'a human mind is a wandering mind, and a wandering mind is an unhappy mind. The ability to think about what is not happening is a cognitive achievement that comes at an emotional cost.'

If you stop to reflect on those moments you experience after a hard workout at the gym or a rigorous jog, they are moments of awareness. The same goes for experiences of fear, such as when you are facing a personal challenge you are especially anxious about – perhaps public speaking. In these moments we are focused, we are present, and we are engaged with an intensity rarely achieved otherwise. Other less desirable negative events, such as rejection, failure or an injury, may also provide these kinds of experiences. Of course, we would not purposely seek these out, as there are other less damaging ways in which we can achieve the same outcomes. Still, even when we do not choose to suffer, we will still experience this increased clarity of awareness, a benefit we may do well to remember.

Although you might prefer the quiet stillness of mindfulness meditation to the sudden jolt of pain, it is worth remembering that pain is not always the worst of all possible outcomes. Across a series of studies Timothy Wilson from the University of Virginia was interested in whether people might find the experience of having nothing to do but sit and think to be enjoyable or unpleasant.[24] In one especially compelling study, the researchers told volunteers there were two parts to the study and that in the first part they were interested in how people rate external stimuli. The volunteers then had electrodes attached to their ankles and were told they could give themselves an electric shock when prompted by the computer. The volunteers were left alone in the room and then rated some pictures and sounds, as well as the electric shock which they self-administered when instructed to do so. They were then asked to imagine that the experimenter gave them $5 and that they could pay to experience or *not* to experience any of the stimuli again. Out of the fifty-five participants in the study, forty-two indicated they would be prepared to pay money to ensure they would not experience the electric shock again.

The volunteers were then introduced to the second part of the study. They were informed they would be sitting by themselves in a room for 10–20 minutes with nothing to do except to entertain themselves with their thoughts. They were also told that they would be given the opportunity to experience one of the stimuli they had rated in the first part of the study while they were sitting there. The volunteers were then asked to wait a few moments while the computer selected which stimulus would be available to them. In fact, all the volunteers were told they would have the option of experiencing the electric shock again while sitting alone in the room.

Researchers found that 71 per cent of men and 26 per cent of women voluntarily shocked themselves at least once while in the room. When they focused on those forty-two participants who, in the first part of the study, reported they would pay money to avoid the electric shocks in the future, 64 per cent of the men and 15 per cent of the women still voluntarily gave themselves at least one electric shock during the thinking period. Experiencing pain is a more desirable option than boredom and a wandering mind.

When we face our limits, when we push ourselves beyond our

comfort zone and to the edge of what it is we can handle, we become more connected to the moment. These challenges focus our minds, sharpen our awareness, and heighten our senses, ensuring that our minds are not preoccupied with unnecessary worries and concerns. It is in these moments that we transcend ourselves, finding a sense of clarity rarely experienced outside intense meditation.

7

The Meaningful Life

You will have bad times, but they will always wake you up to
the stuff you weren't paying attention to.

Robin Williams

Gestalt psychology was developed in Berlin in the late 1800s. It was
based on the premise that people are natural meaning makers; they
are motivated to make sense of the world they live in. In this way,
people can transform the chaos of their worlds into cogent wholes
and use these perceptions to guide their thoughts and behaviour. The
well-known saying 'the whole is greater than the sum of its parts'
comes from Gestalt psychology, only its more correct translation is
'the whole is *other* than the sum of its parts'. The quote comes from
Kurt Koffka, one of the founders of Gestalt psychology, who was
adamant that what we see as the whole is not necessarily greater – it
is just not the same thing. By creating coherence in the world we are
not adding to it but just seeing it in a different, more meaningful,
kind of way.

The Gestalt tendency to create meaning and coherence is exhibited
in the images opposite. When you look at images 1 and 2 your mind
tends to fill in the blanks so to speak; you see a white triangle or the
panda bear. Of course, the triangle does not actually exist and the
panda emerges from a collection of shapes, but your mind fills in
the blanks. What meaning we attribute to the image also depends on
our perspective, a fact that is well illustrated in image 3, where you
might initially see two faces looking at each other or an hour glass,
and these images can flip back and forth from one moment to the

next. Image 4 further illustrates this tendency; you might see a young woman turning away or you might see an old women's profile. Depending on which one appears to you first, it can be difficult to adjust your perspective to see the other.

These simple images demonstrate that our minds are well versed in creating meaning from just a few, often abstract, cues. We are natural meaning makers; we fill in the gaps and try and find coherence even when it is not strictly there. By creating meaning and purpose our minds help us to navigate our environments. If the world were perceived as arbitrary it would be very hard to know what we should do with ourselves. We need order and predictability, and sometimes it is easier to assume things are predictable even when they are not. As the novelist Ann Aguirre wrote in *Outpost*, 'People try to make sense of things, and if they don't know the answers they make them up, because for some, a wrong answer is better than none.'

The other reason we seek out meaning is because it makes us happy. Having purpose and a sense of coherence to our experiences is fundamental to our well-being. When life seems random or unpredictable it leaves us unsettled and unsure. It is for this reason that many forms of psychotherapy, such as Logotherapy or Existential therapy, focus on the development of meaning and the process of sense-making. The idea that making sense of the world and having a purpose in life is a core element of our well-being and happiness aligns with the ancient Greek concept of *eudaimonia*. This can be more or less translated as 'the meaningful life', and is quite different to happiness, or *euphoria*. In many cases, having a meaningful life is not related to experiencing a lot of pleasure or happiness. People who feel their lives are meaningful are not always those who experience the most happiness day-to-day but, as many researchers have noted, they are people who experience high levels of autonomy, personal growth, self-acceptance, mastery and positive interpersonal relationships.[1]

Consider studies showing that people who have children are sometimes less happy than those who do not, but they report having a greater sense of purpose.[2] Bringing children into the world involves a great deal of hard work, but very quickly they provide a new reason for doing the things we do. Whether it be working, trying to stay healthy, or achieving a good work – life balance, these things are suddenly done not only for yourself but for your children, and in this way these goals acquire greater depth.

I often look back on the days prior to parenthood when I would spend my time doing exactly what I wanted. Weekends would be spent eating good food, catching up with friends, enjoying a night out and coming home whenever we wanted to, rather than when the babysitter's time was up. I also often look at my friends who have chosen not to have kids, and their wonderful lifestyles, with just a little tinge of envy – it is nice to do exactly what you want when you want and to have the resources to go about it. Now I do what my kids want when they want it, and my resources are mostly tied up in swimming or piano lessons, school fees, or a number of things that have nothing to do with my own pleasure. Yet, just like most parents, I would not change this for the world. In the larger scheme of things my kids bring me great pleasure, and sometimes I think the harder I must work and the more we have to struggle to get by, the more pleasure my family brings to me. It is not the kind of pleasure I might get from a holiday *sans* children in the Maldives, just sipping cocktails and lazing in the sun, but it *is* meaningful.

This distinction between euphoria (happiness) and eudaimonia (meaning) has been explored in countless studies. Roy Baumeister, now at the University of Queensland, and his colleagues surveyed four hundred adults in the United States to determine what factors contributed to making their lives both happier and more meaningful.[3] They were especially interested in understanding what kinds of things differentiate happiness from meaningfulness. That is, what uniquely contributes to the happy life compared to the meaningful life? Among other things, they found that happier people were more likely to report leading easy lives, to be in good health, to experience more positive emotions, and to be able to buy what they needed without financial strain. People who felt their lives were meaningful but were less happy, on the other hand, had quite different experiences. These individuals tended to sacrifice personal pleasures, engaging instead in difficult undertakings to make substantial contributions to society. They were more likely to experience worry, stress and anxiety, and they also reported more arguments. They engaged in self-reflection, imagining future events and reflecting on past struggles and challenges. They also felt that, compared to others, they had more unpleasant experiences in life. Critically, these were also people who gave more than they took from others, spending more time and money on other people. In contrast, those who were happier but had little meaning in their lives tended to be takers rather than givers, suggesting that while they were relatively carefree, lacking any significant worries and anxieties, they were also more self-absorbed.

Leading a meaningful life is much harder to achieve than leading a happy life. It involves more effort, worry and strain. To make sense of why it is worthwhile, it is helpful to think of 'the good life' in contrast to the happy life. When we think of the good life poolside cocktail-sipping may come to mind, but Stephan Schueller and Martin Seligman from the University of Pennsylvania wondered if we might in fact need more. They surveyed more than 13,000 people, asking them about their 'orientation' to happiness.[4] That is, whether they tended to pursue happiness by seeking out pleasure (e.g. 'I go out of my way to feel euphoric'), engagement (e.g. 'I seek out situations that challenge my skills and abilities') or meaning (e.g. 'In choosing what I do, I always take into account whether it will benefit other people'). The researchers then asked the volunteers to indicate whether they considered themselves happy people, whether they were satisfied with how their lives were going, how many positive and negative emotions they felt on an average day, and their current levels of depression. To get a more objective measure of well-being, they also asked people to indicate their educational and occupational achievements. What they found was that individuals who try to find happiness by seeking out engagement and meaning in their lives tended to report they were more satisfied with their lives, were happier and less depressed, and they also reported more academic and occupational success compared to those who tended to seek happiness by maximizing pleasure. In fact, the more that people sought happiness by maximizing pleasure, the less they tended to experience academic and occupational success in life. Researchers suggested this was because seeking out engagement and meaning in life, as opposed to pleasure, tends to build psychological resources and social connection with others – both of which provide a more balanced and stable sense of well-being.

Traditionally, the relationship between pain and meaning is understood from the perspective that we can cope better with meaningful pain – when we understand our pain it gives it a sense of predictability and purpose that allows us to cope with it better. Pain that is random and meaningless tends to hurt us more. However, this is not the link between pain and meaning that is most important here. Rather, pain itself can be understood as an important trigger for enhancing our perceived meaning in life – sometimes it is *because* of painful experiences that we find meaning and purpose.

FINDING MEANING

An especially powerful study on the link between pain and meaning comes from researchers at the University of British Columbia.[5] There were two relevant elements to this work. The first is that the researchers wanted to know whether common analgesics might not only reduce physical pain, but might also ameliorate existential threat. Building from the study mentioned earlier in Chapter 3 showing acetaminophen could reduce social pain, these researchers wanted to know whether the effects of this common analgesic might also extend to other threatening experiences. As with the link between physical and social pain, they reasoned that due to the overlap in brain regions associated with pain, threat and negative emotion more generally,[6] analgesics could reduce responsiveness to a very broad range of negative experiences. This approach to pain is similar to that taken in this book, and provides yet more evidence that our negative experiences share many important commonalities and are governed by similar neural systems.

The second element of note relates to how researchers demonstrated this effect. Specifically, they attempted to show that acetaminophen would reduce the extent to which people try to bolster a sense of meaning and coherence in their worlds when exposed to existential threat. By using painkillers to ameliorate the threat response, and in turn a search for meaning and coherence, the design of this study can show that negative experiences do in fact motivate such efforts.

Across two studies, researchers randomly assigned participants to receive either 1,000 mg of fast-acting acetaminophen or a 1,000 mg sugar pill placebo. In the first study participants then wrote about their own death – what would happen to their bodies when they died – or wrote an essay about something unpleasant but which was not existentially threatening as a control. In the second study, half the participants watched a 4-minute clip from the short film *Rabbits* by David Lynch, a surrealist movie producer renowned for his ability to disturb, offend or mystify audiences, often leaving them with a sense of anxiety or unease. The other half watched a 4-minute clip of *The Simpsons*. Previous research has shown that when people are confronted with existential threats (such as thoughts of death) they tend to make harsher judgements of others' wrongdoings in an effort to reinforce their own cultural worldview (our cultural worldview protects against these kinds of threats – just think of the number of people who turn to religion when confronted by suffering or death). What researchers found was that, consistent with the previous research, participants who received the placebo pill and experienced the existential threat (the death essay or David Lynch film) tended to make harsher judgements of another person's wrongdoing. Specifically, they recommended a higher bail amount (on a scale between $0 and $999) when asked to imagine a prostitute had been arrested, and they recommended stronger punishment for ice hockey fans when thinking about an actual riot that had recently taken place. This effect, was, however, eradicated for those who were given acetaminophen. The painkiller reduced the extent to which participants had tried to re-establish a sense of justice, because they felt less threatened in the first place.

This study demonstrates that our negative and threatening experiences in life, even those that remind us of death, trigger a motivation to create more meaning in the world. When people are threatened they not only turn to others but also to their belief systems because these worldviews function to make the world coherent and understandable. Just as our minds fill in the gaps in the pictures above (page 149), our belief systems fill in the gaps when explaining why things happen and what our purpose is on the planet – they shape the randomness of the world into a meaningful whole.

The fact is we are more likely to ask 'why' questions when we experience pain compared to when we experience pleasure. When bad things happen we want to know what caused them. This is perhaps why people more often invoke the 'will of God' when it comes to negative rather than positive events in their lives, an observation confirmed in the scatterplot in Figure 11 below. When Kurt Gray and Dan Wegner reversed the scores on a popular health index in the United States and correlated this with people's belief in God, they found that states where people suffer more – such as in Mississippi or Louisiana – also tend to be states where a high proportion of people report a belief in God.[7] This is consistent with the findings of a study investigating the effects of the Christchurch earthquake in 2011 on religious belief.[8] Researchers found that although religious faith in New Zealand overall was in decline between 2009 and 2011, in the earthquake-affected area it increased. A belief in God helps us to make sense of our pain, perhaps because we believe he might relieve it, or because we need someone to blame. Blaming someone also helps us to make sense of what is happening; it provides coherence and restores perceived control (even if only God's control).

There is a great deal of evidence that simply being reminded of death can lead to an increased tendency to believe in God. In one study, researchers from the University of British Columbia asked student volunteers to write either about their own death or about nice foods they enjoy.[9] They were then asked two questions: 'How religious are you?' and 'How strongly do you believe in God?' Volunteers who had written about their own death indicated they were more religious and had a stronger belief in God compared to those who had written about nice food. The researchers ran several other studies using similar death reminders and were able to demonstrate it was not just those who identified as Christian or who were observant in other ways – even non-religious people increased their belief in God. Furthermore, they also showed that Christians were more likely to endorse a belief in Buddha when they were reminded of their own mortality. It isn't so much a case of searching for *the* truth as it is about finding *a* truth.

Further evidence for this link between threat and the motivation to re-establish meaning was reported in the journal *Science* by Jennifer

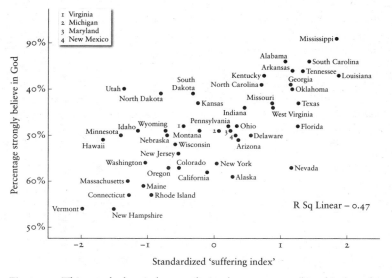

Figure 11. This graph shows the correlation between a combined index of suffering (i.e. poor overall health) and the tendency to report a belief in God. This shows a positive relationship: the more people suffer the more they report believing in God.

Whitson from the University of Texas and Adam Galinsky, now at Columbia Business School.[10] In their studies they specifically focused on the experience of lacking control and how this can lead people to see illusory patterns in the world around them. Feeling out of control is an aversive state that activates the same areas of the brain involved in the experience of fear. It is an experience that, like physical pain, people try to avoid and escape from; when people feel they have lost control they quickly try to re-establish it.

Across several studies researchers induced people to feel they lacked control using different approaches. In some studies, they either asked people to write in vivid detail about a time when they lacked control, or had them complete a simple task where they received feedback that was random and did not align with their actual performance. In one of the studies they then gave people twenty-four pictures to look at, in half of which there were grainy and only just recognizable images embedded, while in the remainder there was no image embedded at all. Three of these pictures are shown below. Under each the volunteers were asked to write a short description of any image they thought they could see. As you might notice, there is an image in pictures 1 and 3, but there is no image in picture 2.

As predicted, when participants were made to feel they lacked control they were more likely to report seeing an image in the pictures where there was no image – they saw an illusory pattern. This aversive experience of not feeling in charge of one's own destiny led

Write one or two words to describe each picture.

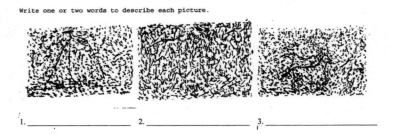

1. _____   2. _____   3. _____

the participants to try to re-establish control by seeing patterns in the world even where they did not exist.

A very similar finding was reported by a Russian group of researchers.[11] They tested five parachute jumpers at various stages in their preparations leading up to the point at which they were to jump out of the plane. They gave the parachutists images that contained random distortions. Embedded within some of the images were numbers, while others had no numbers at all. Consistent with the Yerkes–Dodson Law (which states that a moderate amount of stress is associated with a maximum level of performance), they found that as the parachutists' stress increased so did their performance. As the time to jump got closer, however, and they became more and more stressed, they started to make more errors. To be specific, these errors were false-positives: they were more likely to report seeing numbers in the images when there were none there. Just as the participants above were more likely to think they saw pictures hidden in an image, the parachutists were more likely to think they had seen a number. They saw illusory patterns because they were motivated to find meaning and coherence in the world before launching themselves out of a plane thousands of metres in the air.

To the extent that meaning is the process of making connections between things and seeing predictable patterns, these studies build on the above findings. They show that existential threats, such as a loss of control in life or the possibility of dying while jumping from a plane, lead people to scramble for some sense of coherence. They effectively try to insert meaning where it does not exist, because doing so serves a palliative function – it makes us feel better. Under these circumstances, we are more likely to believe in God (any god for that matter), to vocalize our faith, and to think we can see patterns in random images.

Shigehiro Oishi from the University of Virginia and Ed Diener from the University of Illinois also found a link between self-reported meaning in life and the experience of suffering.[12] They examined data from the Gallup World Poll. This is a massive data set that is freely available and includes many countries. By focusing on specific variables contained within the data, these two academics could look at the relationship between how satisfied people are with their lives, how much meaning they have in their lives, and how these two things are related to the national Gross Domestic Product of 132 countries. To assess life satisfaction, the Gallup survey asked people to indicate where they believed they sat on a ladder scale ranging from 0 (worst possible life) to 10 (best possible life). To assess meaning, the survey asked people to indicate either yes or no to the question 'Do you feel your life has an important purpose or meaning?'

For the first part, the researchers found (unsurprisingly) that satisfaction in life is substantially higher in wealthier nations – people feel they are closer to the top of the ladder and therefore closer to the 'best possible life' when they have more money and resources.

This was, however, negatively correlated to meaning. People in poorer nations reported having more meaning in their lives. It is

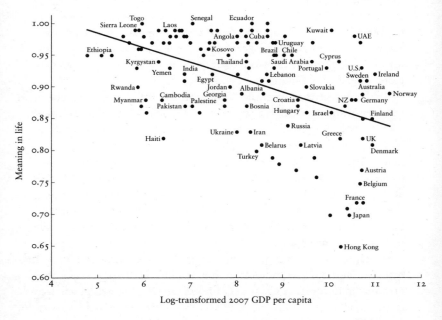

Figure 12. This scatterplot shows the relationship between Gross Domestic Product (GDP) and how much meaning people report they have in their lives. As GDP increases, self-rated meaning in life decreases.

rather amazing to note that between 95 and 100 per cent of the respondents from poverty-stricken Sierra Leone, Togo, Kyrgyzstan, Chad and Ethiopia reported leading meaningful lives. This can be compared to wealthier nations such as Japan, France and Spain, where only two-thirds of the respondents believed their lives had meaning. Researchers concluded that it was the levels of religiosity in these countries that explained higher levels of meaning. People who suffer from lower levels of physical, emotional and social well-being are more likely to believe in God: poor health, fewer resources and inadequate living conditions motivated these individuals to construct meaning in their lives.

Having faith in God out of desperation is clearly not a place most of us would choose to be. In fact believing in God at all may not be our chosen pathway to the meaningful life. Just so, few of us would choose to live in Sierra Leone instead of Norway, Denmark or the UK. The pattern of relationships is, however, what is especially illuminating here. Suffering increases our motivation to search for meaning, and this should be the case even when that suffering is less severe or overwhelming – and meaning need not necessarily come from a belief in God. From a psychological perspective, such beliefs mostly provide reassurance that there is order and predictability in the world and this allows for a sense of well-being. Through this avenue adversity builds a sense of purpose into our lives, and more so compared to when we don't suffer at all.

Although meaning in life may be distinct from feelings of satisfaction and happiness, as the above study shows, this is not to suggest that in order to have a meaningful life we must also be less happy. In fact, it is quite the opposite. The term eudaimonia refers to the meaningful life, but it also refers to the *good* life. While this is distinct from a mere hedonic sense of happiness (euphoria), or a sense that we are satisfied with our lives (as in the study above), it is also most likely to provide a deep sense of joy. As the overarching theme of this book conveys, pleasure can provide positive feelings but they tend to be fleeting and lack a great deal of substance. It is the deeper happiness such as that achieved through our challenges and struggles in life that also provides for a sense of meaning. In fact, the difference between *joy* and *happiness* is useful here. We may say we are happy when things go our way, but we tend to reserve joy for those things in life that also provide a sense of purpose. Indeed, we often say our children bring us joy, even though the research suggests they can make us less happy due to the daily struggles associated with rearing them.

THE POTENCY OF PAIN

Painful events can motivate us to search for and construct more meaning in our lives. Yet, in some cases, negative events themselves tend to be more meaningful than positive ones – they are of more value. Of course, we are unlikely to value our suffering when it has negative outcomes – we certainly tend to value falling in love more than being dumped – but voluntarily exposing ourselves to pain in contexts that allow us to grow and develop can provide a sense that these events are more potent in our lives.

A good example here is travel. People often choose to go on solitary adventures, where they willingly submit themselves to separation, loneliness and hardship. But these adventures also tend to be self-defining moments. This is reflected in the experience of David Zammit, an adventurer and photographer from Malta. In a recent TEDx talk he tells his story of the thirty-two days he spent riding his bicycle across a 1,100-km stretch of desert in outback Australia – the Nullarbor Plain.[13] On day fifteen he recalls being about 140 km away from any point of civilization and realizing he was dangerously low on food and water. He was severely fatigued and starting to experience the occasional hallucination. He recalls that right then and there he felt as if he was close to his own death. Then, as he shows the audience a picture of his bicycle lying on a road in the middle of nowhere, he recounts the expression he had on his face when taking the photo – 'one of pure joy'. He tells how the harsh and unforgiving conditions brought him a sense of tranquillity and peace of mind, and how after gruelling days in the saddle he would watch the sunsets, sitting in the middle of nowhere, with a deep sense of awe, often on the verge of tears. Although Zammit struggles to put everything he experienced into words, to make sense of it all, he recognizes that this raw and authentic experience provided a deep sense of satisfaction. It left him with a pure joy he could not have found through less difficult or challenging pursuits.

If one steps back from these types of adventure-seeking, it becomes quite clear that it is in fact only the difficulty of overcoming such challenges that gives them any appeal in the first place. Riding across the desert or backpacking solo really have few concrete benefits for anyone. There is little that justifies the pursuit of these experiences except that they provide a powerful sense of satisfaction for those who undertake them. This was something that George Mallory, who died attempting to climb Mount Everest, acutely observed when asked what was the point of mountain climbing:

> People ask me, 'What is the use of climbing Mount Everest?' and my answer must at once be, 'It is of no use'. There is not the slightest prospect of any gain whatsoever. Oh, we may learn a little about the behaviour of the human body at high altitudes, and possibly medical men may turn our observation to some account for the purposes of aviation. But otherwise nothing will come of it. We shall not bring back a single bit of gold or silver, not a gem, nor any coal or iron . . . If you cannot understand that there is something in man which responds to the challenge of this mountain and goes out to meet it, that the struggle is the struggle of life itself upward and forever upward, then you won't see why we go. What we get from this adventure is just sheer joy. And joy is, after all, the end of life. We do not live to eat and make money. We eat and make money to be able to live. That is what life means and what life is for.[14]

Mallory really captured the idea that by rising to meet our challenges we can develop a deep sense of purpose – so much so that we often pursue these experiences for no other reason than that they can bring us joy. Note here that both Mallory and Zammit refer to joy rather than happiness, because their experience of overcoming these great challenges provided them with much more than a series of positive feelings; rather, they bestowed on them a deeper sense of well-being. Of course, it is important to reflect on the fact that we don't have to put our lives at risk climbing mountains to get this kind of experience. It can also occur in very minimal kinds of ways, albeit to a lesser extent.

ADDING VALUE

You may recall that in 2014 a drive to raise funds for a disease known as amyotrophic lateral sclerosis (ALS, also known as motor neurone disease) went viral. It became known as the ALS ice-bucket challenge. People could either pour a bucket of ice-water over their heads or donate money to the cause. If someone was challenged they either had to provide video evidence of dousing themselves in ice-cold water, or make a donation. Many people did both and the ALS ice-bucket challenge was a tremendous success. After it took off, visits to a Wikipedia article on ALS went from 163,300 hits per month to 2.89 million. In just one month, between 29 July and 21 August 2014 the ALS association received more than double the donations it had received in the whole of the previous year, with more than 739,000 new donors. By September, donations had exceeded a phenomenal $100 million.

What made the ALS ice-bucket challenge so successful was pain. The spectacle of people pouring ice-water over their heads set off a worldwide phenomenon that had people reaching into their pockets in an unprecedented manner. If those who support a cause are willing to endure a dousing of ice-water, it must be worthy of our time and money. Consider for a moment if the challenge had been to receive a foot massage, would it have worked in the same way? What if it was a bucket of confetti rather than cold water? Sure, it was perhaps funnier to see people endure the shock of a cold drenching, but the act also lent significance to the cause.

So even mild and mostly fun pain can make something seem more worthwhile. The ALS challenge also did something else: it provided a particularly compelling validation of research findings from two psychologists at the University of Warwick and Princeton University. In a series of studies published the year before the ALS challenge went viral, they provided evidence that the experience of pain can increase the value of charitable causes.[15] In one study, they asked university student volunteers to imagine a fundraising event in which they would participate and raise money for tsunami victims. Half of the participants imagined that the fundraiser would be a charity picnic, while the other half imagined it would be a 5-mile charity run. Both groups were told that to attend the event they would need to make a charitable donation. After reading the scenario the volunteers were simply asked how likely they would be to attend the fundraiser and, if they did, how much they would donate to attend. While people were less likely to indicate they would attend the 5-mile run compared to the picnic, those who did say they would attend the fun run on average indicated they would donate $23.87 to do so. This was compared to the picnic-goers, who said they would donate $13.88. Thinking of enduring a 5-mile run made the charity even more meaningful and significant and therefore more worthy of a greater donation.

To replicate these findings in a way that involved actual behaviour rather than imaginary scenarios, researchers ran a second study. This time they had volunteers play a game where they were given $5 and could allocate any portion of it to a public pool of money. This is commonly referred to as a public goods game, where the public pool of money is doubled and split equally among all players. The tension in the game is between contributing to the pool, which benefits everyone, and keeping your own money, which protects your own interests. Those who contribute more (the co-operators) tend to lose out when others do not contribute very much (the free-riders), but if everyone co-operates everyone gets more.

In the study the volunteers entered a large room in groups of three to five and were, one-by-one, led by the experimenter to another, smaller room in a separate part of the building. Here, the experimenter explained the rules of the game, including any costs of contributing money to the public pool, after which the experimenter gave the person a sheet on which to indicate how much of their $5 budget they wanted to allocate to the public pool and how much they wanted to keep for themselves. Volunteers were given as much time as they wanted to make their allocation decisions. In each session, the volunteers were arbitrarily assigned, as a group, to one of two experimental conditions (or game types). The control group played the standard version of the public goods game, as described so far. The other group were given similar instructions but with one key difference: they were told they had to endure an aversive experience if they wanted to allocate any amount of money to the public pool. Specifically, that they had to immerse both of their hands in a bucket of ice-water for 60 seconds. These volunteers were told the result of failing to complete this physical task was that the money they contributed to the public pool would not double in value.

Therefore, the volunteers in the control condition could contribute to the public pool or not, while the volunteers in the pain condition had to withstand pain in order to make their contribution. Those participants had two reasons not to contribute – they could keep their money and they could avoid the experience of pain.

One-by-one the players assigned to the pain group were led into the room with the ice-bucket where they made their choice, either making a 'painful' donation, or keeping their money for themselves. These players donated an average of $4.17 to the public pool, while participants who did not have to withstand the ice-bucket before giving their donation on average gave $3.18. So, the volunteers who had to endure pain to make a contribution donated 83 per cent of their total budget, whereas the volunteers who did not have to endure anything donated 64 per cent. Further, volunteers who endured the ice-bucket test were more likely than those who did not to contribute their entire budget to the public pool (67 per cent vs. 28 per cent).

This experiment demonstrates what on the surface of things seems rather counterintuitive . . . and might still seem counterintuitive, if the predictions had not been borne out so perfectly just a year later with the ALS ice-bucket challenge.

But the most useful question remained: why? To answer this, the researchers conducted another study. They suspected that all this had to do with meaningfulness – the pain that was experienced or expected endowed the act of donating with additional meaning. So they ran the same study, described above, where they asked people to either imagine a charity picnic or a fun run to raise money for victims of war and genocide. This time, however, they also asked the volunteers these additional questions:

1. How meaningful (to you) would this experience be?
2. How meaningful (to you) would your participation in the event be?
3. How meaningful (to you) would your contribution be?

The volunteers responded to each question using a scale ranging from 1 ('Not at all meaningful') to 10 ('Very meaningful').

As in the earlier study, volunteers who imagined the fun run offered to donate more money (£17.95 on average) than volunteers who imagined the picnic (£5.74 on average). Also, the volunteers who imagined the fun run judged the fundraiser to be more meaningful (6.46 on average) than those who imagined the picnic (5.93 on average). Using statistical analysis, the researchers were able to demonstrate it was the perceived meaningfulness of the fundraising event that explained the larger donations. Volunteers gave more when imagining the fun run because it had to them greater inherent worth.

Pain indeed does make things seem more meaningful and, in turn, it makes those things seem more worthy of our time and resources – they become more highly valued. This is largely inconsistent with what many people would predict or expect. Indeed, pleasure tends to be associated with positive value and pain tends to be associated with negative value. Yet, this intuitive idea is perhaps not so accurate after all. At least in some contexts, the experience of pain can engender another level of worth.

DEATH GIVES MEANING TO LIFE

As we all know, scarcity is a primary driver of value. This is why diamonds are so expensive and old coins are worth more than common currency. If something is both desirable and starts to become scarce, its value appreciates. Whether it is things that money can buy or things that we cannot put a price on, they all still follow the same pattern of value appreciation. Take ancient woodlands. We (mostly) cannot buy ancient woodlands, but one reason we revere them is because they are scarce. The same goes for endangered species. In Australia, we tend to value Tasmanian tigers more than kangaroos – not because we like them better, but because there are fewer of them.

This link between scarcity and value also intrigued a group of researchers from the University of Missouri, led by Laura King.[16] They were interested in accounts of individuals who had had brushes with death and emerged from these experiences with a new appreciation for the value of life. It made them wonder whether death reminders might make life seem more scarce and, in this way, increase its perceived value.

To test these ideas the researchers provided volunteers with a word-finding puzzle – much like the ones you might see in a newspaper. The volunteers' task was to find all the words they could. The trick was that for half of the volunteers the hidden words all related to death (e.g. 'dead', 'tombstone', etc.). This meant half of the volunteers were led to focus on death-related words while completing the task, while the others were not.

Next, volunteers responded to several questions about how satisfied they were with their lives and how meaningful they believed their lives were. What the researchers found was that the participants who had searched for death-related words in the word-finding task rated their lives as more satisfying and more meaningful – thinking of death had in effect increased the perceived value of life.

Also interested in this relationship, another team of researchers (also from the University of Missouri) examined the link between death and life in a different way.[17] They wanted to know whether death reminders lead people to search for meaning in their lives, and whether different people may search for this meaning in different ways. Across several studies, they reminded volunteers of death by asking them to respond to the following: 'Please briefly describe the emotions that the thought of your own death arouses in you', and 'Jot down, as specifically as you can, what you think will happen to you once you are physically dead.' Other volunteers wrote about an experience that was unrelated to death, as a control. Both groups then completed a measure of how much meaning they felt their lives had. Specifically, they indicated on a scale from 1 (absolutely untrue) to 7 (absolutely true) whether the following statements were true for them:

1. I understand my life's meaning.
2. My life has a clear sense of purpose.
3. I have a good sense of what makes my life meaningful.
4. I have discovered a satisfying life purpose.
5. My life has no clear purpose (this was reverse scored).

The researchers thought one factor that may affect how people respond to death is whether they have a high need for structure or not – that is, whether they prefer predictability and order in their worlds as opposed to chaos. For these individuals, seeing the world as ordered and predictable is a way of assuaging their fear of death, which also suggests that they might be more likely to see their worlds as coherent and meaningful when reminded of death. This was exactly what they found. People who preferred structure, when they were reminded of death, tended to see their lives as more meaningful, compared to those who had not written about their own death. That is, after being reminded of death they were more likely to say their lives had a clear sense of purpose and they had a good sense of what makes their lives coherent.

They also found, however, that people who did not have this same tendency to prefer order over chaos tended to see their lives as *less* meaningful after being reminded of death. This seemed puzzling. Is it that some people really see their lives as lacking in meaning and purpose when reminded of death? The researchers dug a little deeper, aiming to understand whether this was in fact true, or whether these individuals might seek meaning in different ways.

To this end the researchers conducted another study. They again asked half of the volunteers to write about their own death. They then asked them to respond to the questions below, which are designed to capture how much people like to explore novel ideas, have novel experiences and meet novel people:

1. I would like to take a class that is unrelated to my major just because it interests me.
2. I would like to try bungee jumping, skydiving, or other adventurous activities.
3. If I had the time and money, I would like to travel overseas this summer.
4. I would like to explore someplace that I have never been before.
5. I would like to have several friends who are very different from each other.
6. I would like to spend a semester studying abroad.

7. I would like a job that was unusual and different.
8. I would like to have the chance to meet strangers.
9. If given the chance, I would enjoy exploring novel ideas or theories.
10. I would like to explore the woods and interesting places near my town.
11. I would enjoy being introduced to new people.
12. I would pick up a book on an interesting topic and read some of it.
13. If I had time, I would enjoy watching TV shows on interesting topics such as science, history, art, or culture.
14. I would like to explore the ideas of foreign cultures.
15. I would enjoy joining a student group composed of a wide range of people that I don't know.
16. I would like to go to a modern art museum.
17. I would like to strike up a conversation with a stranger on a bus or airplane and open up to the person.
18. I would like to go to a party if I didn't know very many of the people.

For people who did not have a strong preference for structure, reminders of their own death motivated them to seek out novel experiences and open their minds (that is, they endorsed the above items to a greater extent). This suggests that while people who like structure tend to see their lives as more meaningful in response to being reminded of death, people who do not prefer order and structure tend to see their lives as less meaningful, but instead they express a desire to seek out novel experiences – a kind of bucket-list response to death reminders, if you like.

Could this search for novelty be a way in which these individuals try to find meaning in their lives after being reminded of death? This is what the researchers examined next. They again asked half of their volunteers to write about their own death. After doing this, they asked the volunteers to rank six different topics in terms of their familiarity to them. They included three topics that were likely to be familiar to most of the people (current trends in pop culture, a typical day for American college students, and inside a shopping mall) and three topics that were likely to be unfamiliar (living with the Aborigines of Australia, the hidden meaning of abstract art, and the customs and rituals of the Iranian people). The researchers next asked half of their volunteers to imagine that they were conducting an internet search on the topic they ranked as most familiar, while they asked the other half of their volunteers to imagine searching the topic they ranked as least familiar. Finally, they asked the volunteers to think about the important information that the search would yield, to write down what they thought would be the three most important pieces of information resulting from the search, and visualize themselves reading over that information. The volunteers then responded to the questions above (page 166) about their perception of their life's meaning.

When reminded of death people who do not need structure try to establish more meaning in their lives by seeking out novel experiences. The participants who had been reminded of death and then imagined researching on the internet a topic that was novel and unfamiliar to them rated their lives as more meaningful on the same scale as above; they saw their lives as having more purpose.

People respond to death in different ways. Some turn to the things they know, such as religion or other knowledge structures that give their lives a clear sense of coherence and purpose. Others, however, who are not naturally drawn to these kinds of meaning systems, seek out novelty – they want to meet new people, have new experiences, experience new art, or learn about foreign cultures – and it is in this way that they feel their lives have depth.

THE STUFF YOU WEREN'T PAYING ATTENTION TO

On 17 January 1994 the Northridge earthquake – measured at 6.7 on the Richter scale – took place in the Los Angeles area, resulting in fifty-seven deaths. A group of researchers from the University of Kentucky and the University of California surveyed employees of the Sepulveda Veterans Affairs Medical Center which was in one of the worst-affected areas. The surveys occurred at two time points, approximately one month after the earthquake, and again approximately a month later.

In the first survey the employees were asked, 'At any time during the earthquake did you think you might die?' In the second survey they were asked to indicate how much importance they placed on different goals in life. Some of these were extrinsic goals, such as being attractive, receiving acknowledgement and praise, having possessions, and advancing their careers. Others were intrinsic goals such as cultivating close friendships, giving and receiving love, developing as a person, and doing creative work. There is a very large body of work within psychology showing that pursuing intrinsic goals in life is more likely to make us happier and healthier.[18] The employees were asked to indicate how important these goals were to them before the earthquake occurred, and then to indicate how important they were to them currently.

The surveys revealed that people's priorities shifted from giving importance to extrinsic goals before the earthquake to placing more importance on intrinsic goals, and this was still evident even several months after the earthquake. Most importantly, this shift in emphasis was predicted by the extent to which the employees reported having thoughts of dying during the earthquake (as measured in the first survey). Being aware of death, and even feeling threatened by death, is certainly not enjoyable or recommended – however, it would seem beneficial in reaffirming our deeper-rooted values. The more we consciously engage with our own mortality the more likely we are to focus on things that matter; to seek out things that are ultimately likely to provide more depth to our lives.

Conclusion

We are in the midst of a medication epidemic. The use of painkillers is sharply on the rise, as is reliance on sleeping pills, beta-blockers for anxiety, and antidepressants. What is utterly mystifying is that we are increasingly seeking to numb ourselves to our daily discomforts in a world where our levels of comfort are at an all-time high. As a species, we have overcome the challenges of our environment in ways that no other animal has. We started by developing basic tools from sticks and stones and now have access to instant information via our iPhones, and driverless cars. We harnessed the power of fire to cook our food and progressed to Michelin-starred restaurants and fine wine. We outsmarted our predators and proceeded to place them in zoos.

Just as we have attempted to conquer our external environment, we have also attempted to conquer our internal worlds. Our progress towards this end has evolved exponentially in the last hundred or so years. With the advent of modern analgesics, we suddenly found ourselves in a position where many of us could take control of our own pain, or at least extend our control far beyond where it was previously. We now find ourselves in a world where futurists such as David Pearce feel confident in making the prediction that the world's last unpleasant experience will be a dateable event.

So why, then, in a world where a life free of pain appears to be within our collective grasp, are we experiencing an increase in pain-related issues, in pain clinics, in a reliance on pain-related medication, in depression, anxiety, and the use of mood stabilizers and stress reducers? Why is our utopian vision turning on us, and causing more pain than it has resolved? Why is the burden of pain, depression and anxiety on the rise just at a time when we appear to be gaining the upper hand?

The answer to this puzzle can be found in our fundamental orientation to pain. We have come to see it as a threat, as something we should avoid at all costs, although the faster we run from it the more painful it gets. As the title of a recent post on the popular website Psychology Today asks, 'Is our aversion to pain killing us?'[1] In the case of Heath Ledger and Whitney Huston, both of whom died from an overdose of pain medication or sleeping pills, and many other addicts, the answer is yes it is, literally.

A 2016 report from the Centers for Disease Control in the United States indicates the number of deaths from prescription opioids (painkillers) is on the rise,[2] and it has been fuelled by a four-fold increase in the prescription of opioid pain relievers over the past twenty years within America. This same rise is evident in the use of mental health medications. According to *America's State of Mind* Report,[3] one in five adults were using some form of mental health medication in 2010, up 22 per cent since 2001.

We are both reporting and treating more pain, but it is not our levels of pain that are getting worse, it is our subjective threshold that has changed – we find even our mild pains intolerable. We expect never to be in pain, and when we are it represents a tear not only in our experience of pleasure, but in the very fabric of our realities. We believe our lives are *supposed* to be free of pain, and from this perspective painful experiences are not only physically and emotionally threatening: they are existentially compromising.

As we numb, run and hide from pain, we are not only giving too much credit to that pain, we are weakening our capacity to adaptively manage it in the future. Painful experiences are important for our development. We know it is through adversity that we grow, become stronger and respond better to future challenges. This occurs not only at the psychological level but at the biological level – our minds and our bodies deal with stress better when they have been exposed to it before. We vaccinate our children because we know it will make them resistant to disease, but we forget that pain and stress work in the same ways – they provide a psychological vaccination against future pain and stress. By avoiding pain, we are making ourselves and our children emotionally unstable, more fearful of discomfort, and more likely to seek out the salve of painkillers or antidepressants for even mild upsets.

So what do we do now?

One thing we can do is change how we think. By engaging with a different perspective on pain, and understanding that it is central to many desirable experiences in life, we can begin to relate to it in different ways. Whether it is the challenge of completing a marathon, the joy of children, the reward of hard work, or the cost of helping others, these experiences bring us immense joy. By recognizing it is not simply our pleasures that make these experiences rewarding but our pains which play a necessary and important role, we can see more clearly what truly brings us happiness. In this way, we can begin to understand that our pains are valuable and important. They are not simply experiences that we would be better off without. By seeing value in our negative experiences, we are also better able to respond to them. Whether it is the challenges we choose or the adversities thrust upon us, we can see a different side to these experiences, one that allows us to embrace them rather than run from them. It is this orientation to pain that will ensure not only that we expose ourselves to experiences where we learn and grow, but also that we are better able to cope with the various tests life throws at us. How we respond to our painful experiences fundamentally shapes how those experiences impact on our lives. By developing a more open, understanding and accepting orientation to pain we are better equipped to respond with courage.

Secondly, we can change how we behave. We could begin to take a few more calculated risks in life, and even let our children do the same. Engaging with risk is important for mental health, and our fear of misfortune fosters an overly protectionist mind-set. By removing the cotton wool we can open a new, more exciting approach to living, one that sometimes fails, and sometimes hurts, but one that also provides us with a sense of achievement. We could also change how we seek out pleasure. Rather than searching for comfort, we might begin to actively search for discomfort: pushing our bodies rather than resting them, spending our holidays seeking novel and challenging experiences rather than luxurious ones. Bound up in this more uncomfortable approach to living is the limitless potential for learning and personal development. We might also begin to embrace our feelings of sorrow or loss, and do the same when others come to us with such concerns. By spending time with this uncomfortable emotional content we are not wasting time, but gaining insight and understanding that could not otherwise be achieved.

By valuing and engaging with these uncomfortable experiences, we advance our capacity for navigating life in a way that ultimately produces more pleasure. Through this, we will come to understand that our capacity for pleasure is largely dependent on our ability to endure pain. This insight was also shared by Leonardo da Vinci in his

Figure 13. 'Allegories of Pleasure and Pain' (1480) by Leonardo da Vinci

'Allegory of Pleasure and Pain'. In his notebooks, he described the two interconnected allegorical figures as follows:

> Pleasure and pain are represented as twins, as though they were joined together, for there is never one without the other . . . They are made with their backs turned to each other because they are contrary the one to the other. They are made growing out of the same trunk because they have one and the same foundation, for the foundation of pleasure is labor with pain, and the foundations of pain are vain and lascivious pleasures.[4]

The poet John Keats also saw pleasure and pain as intimately intertwined. His poems were often built on the notion that there is immense beauty in pain. In a love letter to Fanny Brawne he writes: 'I have had two luxuries to brood over in my walks. Your loveliness and the hour of my death. O that I could have possession of them both in the same minute.' Both Keats and Leonardo see pain and pleasure as deeply interconnected. Both understand that one cannot occur without the other, and for Keats this leads to a longing that he might experience both beauty and pain in the same moment. They go together – her loveliness and his death – and for him they join to create something beautiful; something transcendent.

Our current approach to seeking happiness as an end in itself has bred in us an unhealthy fetish for instant gratification. We have lost the space in our society and in our minds for melancholy, mourning or sorrow. We cast aside our negative experiences as unwanted and unnecessary, and in so doing we undermine life's many forms. We all know that our pains are useful because they warn us of danger or inform us of our limits, but what we have overlooked is that our painful experiences are also central to our psychological well-being. Our pains provide an anchor for our pleasures; they make us empathetic and pro-social, leading to the possibility of greater fulfilment. Our painful experiences are what make us human. They ground our thinking, our purpose, and they ground us in others. In our rush to eradicate pain, we have simply forgotten that it is often in our moments of deepest despair that truth and beauty present themselves with a clarity and intensity we would never otherwise know.

References

INTRODUCTION

1 Shackman, A. J., Salomons, T. V., Slagter, H. A., Fox, A. S., Winter, J. J. and Davidson, R. J. (2011). The integration of negative affect, pain and cognitive control in the cingulate cortex. *Nature Reviews Neuroscience*, 12 (3), 154–67.

2 Eisenberger, N. I. and Lieberman, M. D. (2004). Why rejection hurts: A common neural alarm system for physical and social pain. *Trends in Cognitive Sciences*, 8 (7), 294–300; Panksepp, J. (2003). Feeling the pain of social loss. *Science*, 302 (5643), 237–9.

3 Haslam, N. (2016). Concept creep: Psychology's expanding concepts of harm and pathology. *Psychological Inquiry*, 27 (1), 1–17.

4 Averill, J. R. (1980). On the paucity of positive emotions. In K. Blankstein, P. Pliner and J. Polivy (eds.), *Advances in the Study of Communication and Affect* Vol. 6. New York: Plenum.

5 Rozin, P. and Royzman, E. B. (2001). Negativity bias, negativity dominance, and contagion. *Personality and Social Psychology Review*, 5 (4), 296–320.

1. HAVE WE REACHED PEAK COMFORT?

1 Zanden, J. L. et al. (eds.) (2014), *How Was Life? Global Well-being Since 1820*. Paris: OECD Publishing; http://www.maxroser.com/category/100-charts-project/page/2/.

2 http://www.inc.com/ss/will-yakowicz/10-best-industries-on-2014-inc-5000.html.

3 http://www.maxroser.com/category/100-charts-project/page/2/.

4 http://www.bain.com/bainweb/PDFs/Bain_Worldwide_Luxury_Goods_Report_2014.pdf.

5 Diogenes Laertius, *Lives of Eminent Philosophers*, 10.22 (trans. C. D. Yonge).

6 http://www.apa.org/news/press/releases/2016/08/tinder-self-esteem.aspx.

7 http://www.hedweb.com/.

8 Fredrickson, B. L. (2000). Extracting meaning from past affective experiences: The importance of peaks, ends, and specific emotions. *Cognition & Emotion*, 14 (4), 577–606.

9 Fredrickson, B. L. and Kahneman, D. (1993). Duration neglect in retrospective evaluations of affective episodes. *Journal of Personality and Social Psychology*, 65 (1), 45–55.

10 Durso, G. R., Luttrell, A. and Way, B. M. (2015). Over-the-counter relief from pains and pleasures alike: Acetaminophen blunts evaluation sensitivity to both negative and positive stimuli. *Psychological Science*, 0956797615570366.

11 Schooler, J. W., Ariely, D. and Loewenstein, G. (2003). The pursuit and assessment of happiness can be self-defeating. *The Psychology of Economic Decisions*, 1, 41–70.

12 Beecher, H. K. (1946). Pain in men wounded in battle. *Annals of Surgery*, 123 (1), 96–105.

13 Benedetti, F., Thoen, W., Blanchard, C., Vighetti, S. and Arduino, C. (2013). Pain as a reward: Changing the meaning of pain from negative to positive co-activates opioid and cannabinoid systems. *Pain*, 154, 361–7.

14 Bingel, U., Wanigasekera, V., Wiech, K., Mhuircheartaigh, R. N., Lee, M. C., Ploner, M. and Tracey, I. (2011). The effect of treatment expectation on drug efficacy: Imaging the analgesic benefit of the opioid remifentanil. *Science Translational Medicine*, 3 (70), 70ra14.

15 https://www.accc.gov.au/media-release/court-finds-nurofen-made-mis leading-specific-pain-claims.

16 Kaptchuk, T. J., Stason, W. B., Davis, R. B., Legedza, A. R., Schnyer, R. N., Kerr, C. E., . . . and Goldman, R. H. (2006). Sham device v. inert pill: Randomised controlled trial of two placebo treatments. *British Medical Journal*, 332 (7538), 391–7.

17 Crombez, G., Vlaeyen, J. W., Heuts, P. H. and Lysens, R. (1999). Pain-related fear is more disabling than pain itself: Evidence on the role of pain-related fear in chronic back pain disability. *Pain*, 80 (1), 329–39.

18 Lethem, J., Slade, P. D., Troup, J. D. G. and Bentley, G. (1983). Outline of a fear-avoidance model of exaggerated pain perception – I. *Behaviour Research and Therapy*, 21 (4), 401–8.

19 Harmon-Jones, E., Harmon-Jones, C., Amodio, D. M. and Gable, P. A. (2011). Attitudes toward emotions. *Journal of Personality and Social Psychology*, 101 (6), 1332–50.

20 http://www.bemindful.org/kabatzinnart.htm.

2. THE COTTON WOOL GENERATION

1 Bastian, B., Jetten, J. and Fasoli, F. (2011). Cleansing the soul by hurting the flesh: The guilt-reducing effect of pain. *Psychological Science*, 22, 334–5.

2 Bastian, B., Jetten, J. and Stewart, E. (2013). Physical pain and guilty pleasures. *Social, Psychological and Personality Science*, 4, 215–19.

3 Rojstaczer, S. and Healy, C. (2012). Where A is ordinary: The evolution of American college and university grading, 1940–2009. *Teachers College Record*, 114 (7), 1–23.

4 Baumeister, R. F., Campbell, J. D., Krueger, J. I. and Vohs, K. D. (2003). Does high self-esteem cause better performance, interpersonal success, happiness, or healthier lifestyles? *Psychological Science in the Public Interest*, 4 (1), 1–44.

5 Twenge, J. M. and Campbell, W. K. (2009). *The Narcissism Epidemic: Living in the Age of Entitlement*. New York: Simon and Schuster.

6 Huffington Post (2012). When everyone gets a trophy, no one wins. http://www.huffingtonpost.com/michael-sigman/when-everyone-gets-a-trop_b_1431319.html.

7 Uhls, Y. T. and Greenfield, P. M. (2011). The rise of fame: An historical content analysis. *Cyberpsychology: Journal of Psychosocial Research on Cyberspace*, 5 (1).

8 Uhls, Y. T. and Greenfield, P. M. (2012). The value of fame: Preadolescent perceptions of popular media and their relationship to future aspirations. *Developmental Psychology*, 48 (2), 315–26.

9 Twenge, J. M., Konrath, S., Foster, J. D., Campbell, W. K. and Bushman, B. J. (2008). Egos inflating over time: A cross-temporal meta-analysis of the narcissistic personality inventory. *Journal of Personality*, 76 (4), 875–902.

10 Sayer, L. C., Bianchi, S. M. and Robinson, J. P. (2004). Are parents investing less in children? Trends in mothers' and fathers' time with children. *American Journal of Sociology*, 110, 1–43.

11 The National Center for Safe Routes to School (2011). How Children Get to School: School Travel Patterns from 1969 to 2009. http://

saferoutesinfo.org/sites/default/files/resources/NHTS_school_travel_
report_2011_0.pdf. Accessed 12 April 2012.

12 Pinker, S. (2011). *The Better Angels of Our Nature: Why Violence Has
    Declined.* New York: Viking.

13 Putnam, R. D. (2000). *Bowling Alone: The Collapse and Revival of
    American Community.* New York: Simon and Schuster.

14 Kindlon, D. J. (2003). *Too Much of a Good Thing: Raising Children
    of Character in an Indulgent Age.* New York: Hyperion Books.

15 Sandseter, E. B. H. and Kennair, L. E. O. (2011). Children's risky play
    from an evolutionary perspective: The anti-phobic effects of thrilling
    experiences. *Evolutionary Psychology*, 9 (2), 257–84.

16 Allen, N. B. and Badcock, P. B. (2003). The social risk hypothesis of
    depressed mood: Evolutionary, psychosocial, and neurobiological per-
    spectives. *Psychological Bulletin*, 129 (6), 887.

17 Mata, R., Josef, A. K. and Hertwig, R. (2016). Propensity for risk tak-
    ing across the life span and around the globe. *Psychological Science*,
    0956797615617811.

18 https://www.washingtonpost.com/news/grade-point/wp/2016/03/24/
    someone-wrote-trump-2016-on-emorys-campus-in-chalk-some-students-
    said-they-no-longer-feel-safe/.

19 http://nymag.com/thecut/2016/08/university-of-chicago-bans-trigger-
    warnings-safe-spaces.html.

3. PAINFUL PLEASURES

1 http://www.theage.com.au/victoria/extreme-body-piercers-hooked-on-
   adrenalin-20150405-1mdqof.html.

2 Rozin, P., Guillot, L., Fincher, K., Rozin, A. and Tsukayama, E. (2013).
   Glad to be sad, and other examples of benign masochism. *Judgment
   and Decision Making*, 8 (4), 439–47.

3 Rozin, P. and Kennel, K. (1983). Acquired preferences for piquant
   foods by chimpanzees. *Appetite*, 4, 69–77.

4 http://www.abc.net.au/news/2015-07-22/winter-swimmers-brighton-
   baths/6636306.

5 Kahneman, D., Wakker, P. P. and Sarin, R. (1997). Back to Bentham?
   Explorations of experienced utility. *Quarterly Journal of Economics*,
   112 (2), 375–405.

6 Mellers, B., Schwartz, A. and Ritov, I. (1999). Emotion-based choice.
   Journal of Experimental Psychology: General, 128 (3), 332–45.

7 Leknes, S., Berna, C., Lee, M. C., Snyder, G. D., Biele, G. and Tracey, I. (2013). The importance of context: When relative relief renders pain pleasant. *Pain*, 154 (3), 402–10.

8 Launer, J. (2004). The itch. *Quarterly Journal of Medicine*, 97 (6), 383–4.

9 Solomon, R. L. and Corbit, J. D. (1974). An opponent-process theory of motivation: I. Temporal dynamics of affect. *Psychological Review*, 81 (2), 119–45.

10 Franklin, J. C., Lee, K. M., Hanna, E. K. and Prinstein, M. J. (2013). Feeling worse to feel better: Pain-offset relief simultaneously stimulates positive affect and reduces negative affect. *Psychological Science*, 0956797612458805.

11 Taylor, A. M., Becker, S., Schweinhardt, P. and Cahill, C. (2016). Mesolimbic dopamine signaling in acute and chronic pain: Implications for motivation, analgesia, and addiction. *Pain*, 157 (6), 1194.

12 MacDonald, G. and Leary, M. R. (2005). Why does social exclusion hurt? The relationship between social and physical pain. *Psychological Bulletin*, 131, 202–23.

13 Shackman, A. J., Salomons, T. V., Slagter, H. A., Fox, A. S., Winter, J. J. and Davidson, R. J. (2011). The integration of negative affect, pain and cognitive control in the cingulate cortex. *Nature Reviews Neuroscience*, 12, 154–67.

14 DeWall, C. N., MacDonald, G., Webster, G. D., Masten, C. L., Baumeister, R. F., Powell, C., . . . and Eisenberger, N. I. (2010). Acetaminophen reduces social pain: Behavioral and neural evidence. *Psychological Science*, 21, 931–7.

15 Woo, C.-W., Koban, L., Kross, E., Lindquist, M. A., Banich, M. T., Ruzic, L., . . . and Wager, T. D. (2014). Separate neural representations for physical pain and social rejection. *Nature Communications*, 5. doi:10.1038/ncomms6380; Iannetti, G. D., Salomons, T. V., Moayedi, M., Mouraux, A. and Davis, K. D. (2013). Beyond metaphor: Contrasting mechanisms of social and physical pain. *Trends in Cognitive Sciences*, 17 (8), 371–8.

16 Doyne, E. J., Ossip-Klein, D. J., Bowman, E. D., Osborn, K. M., McDougall-Wilson, I. B. and Neimeyer, R. A. (1987). Running versus weight-lifting in the treatment of depression. *Journal of Consulting and Clinical Psychology*, 55 (5), 748.

17 Borsook, T. K. and MacDonald, G. (2010). Mildly negative social encounters reduce physical pain sensitivity. *Pain*, 151 (2), 372–7.

18 Solomon, R. L. and Corbit, J. D. (1974). An opponent-process theory of motivation: I. Temporal dynamics of affect. *Psychological Review*, 81 (2), 119–45.

19 Fernandes, H. B., Kennair, L. E. O., Hutz, C. S., Natividade, J. C. and Kruger, D. J. (2016). Are negative postcoital emotions a product of evolutionary adaptation? Multinational relationships with sexual strategies, reputation, and mate quality. *Evolutionary Behavioral Sciences*, 10 (4), 219–44.

20 Solomon, R. L. (1980). The opponent-process theory of acquired motivation: The costs of pleasure and the benefits of pain. *American Psychologist*, 35 (8), 691.

21 Richins, M. L. (2013). When wanting is better than having: Materialism, transformation expectations, and product-evoked emotions in the purchase process. *Journal of Consumer Research*, 40 (1), 1–18.

22 Cooney, G., Gilbert, D. T. and Wilson, T. D. (2014). The unforeseen costs of extraordinary experience. *Psychological Science*, 25 (12), 2259–65.

23 Small, D. M., Zatorre, R. J., Dagher, A., Evans, A. C. and Jones-Gotman, M. (2001). Changes in brain activity related to eating chocolate. *Brain*, 124 (9), 1720–33.

4. GETTING TOUGH

1 Mancini, A. D., Littleton, H. L. and Grills, A. E. (2016). Can people benefit from acute stress? Social support, psychological improvement, and resilience after the Virginia Tech campus shootings. *Clinical Psychological Science*, 4 (3), 401–17.

2 Peterson, C. and Seligman, M. E. (2003). Character strengths before and after September 11. *Psychological Science*, 14 (4), 381–4.

3 Morris, D. B. (1991). *The Culture of Pain*. Oakland, CA: University of California Press.

4 Scarry, E. (1985). *The Body in Pain: The Making and Unmaking of the World*. New York: Oxford University Press.

5 Maslow, A. H. (1943). A theory of human motivation. *Psychological Review*, 50 (4), 370–96.

6 Maslow, A. H. (1954). *Motivation and Personality*. New York: Harper, p. 236.

7 Levine, S. (1959). The effects of differential infantile stimulation on emotionality at weaning. *Canadian Journal of Psychology*, 18, 243–7.

8 Holmes, Frances B. (1935). An experimental study of children's fears. In A. T. Jersild and Frances B. Holmes (eds.), *Children's Fears*. New

York: Teachers College, Columbia University (Child Development Monographs 20); Hull, C. L. (1943). *Principles of Behavior*. New York: Appleton.

9   Kobasa, S. C., Maddi, S. R. and Kahn, S. (1982). Hardiness and health: a prospective study. *Journal of Personality and Social Psychology*, 42 (1), 168–77.

10 Neff, L. A. and Broady, E. F. (2011). Stress resilience in early marriage: Can practice make perfect? *Journal of Personality and Social Psychology*, 101, 1050–67.

11 Seery, M. D., Leo, R. J., Lupien, S. P., Kondrak, C. L. and Almonte, J. L. (2013). An upside to adversity? Moderate cumulative lifetime adversity is associated with resilient responses in the face of controlled stressors. *Psychological Science*, 0956797612469210.

12 Seery, M. D., Holman, E. A. and Silver, R. C. (2010). Whatever does not kill us: Cumulative lifetime adversity, vulnerability, and resilience. *Journal of Personality and Social Psychology*, 99 (6), 1025–41.

13 Antonovsky, A. (1987). *Unraveling the Mystery of Health. How People Manage Stress and Stay Well*, San Francisco: Jossey-Bass Publishers.

14 Blascovich, J. and Tomaka, J. (1996). The biopsychosocial model of arousal regulation. *Advances in Experimental Social Psychology*, 28, 1–52.

15 Dienstbier, R. A. (1989). Arousal and physiological toughness: Implications for mental and physical health. *Psychological Review*, 96 (1), 84–100.

16 Overmier, J. B. and Seligman, M. E. (1967). Effects of inescapable shock upon subsequent escape and avoidance responding. *Journal of Comparative and Physiological Psychology*, 63 (1), 28–33.

17 Weiss, J. M., Glazer, H. I., Pohorecky, L. A., Brick, J. and Miller, N. E. (1975). Effects of chronic exposure to stressors on avoidance-escape behavior and on brain norepinephrine. *Psychosomatic Medicine*, 37 (6), 522–34.

18 Johansson, G., Frankenhaeuser, M. and Magnusson, D. (1973). Catecholamine output in school children as related to performance and adjustment. *Scandinavian Journal of Psychology*, 14 (1), 20–28.

19 Ellertsen, B., Johnsen, T. B. and Ursin, H. (1978). Relationship between the hormonal responses to activation and coping. In H. Ursin, E. Baade and S. Levine (eds.), *Psychobiology of Stress: A Study of Coping Men*. New York: Academic Press (pp. 105–24).

20 Ichinohe, T., Pang, I. K., Kumamoto, Y., Peaper, D. R., Ho, J. H., Murray, T. S. and Iwasaki, A. (2011). Microbiota regulates immune defense

against respiratory tract influenza A virus infection. *Proceedings of the National Academy of Sciences*, 108 (13), 5354–9.

21 Solomon, G. S., Kay, N. and Morley, J. E. (1986). Endorphins: A link between penalty, stress, emotions, immunity, and disease? In N. P. Plotnikoff, R. E. Faith, A. J. Murgo and R. A. Good (eds.), *Enkephalins and Endorphins: Stress and the Immune System*. New York: Plenum (pp. 129–44).

5. CONNECTING WITH OTHERS

1 Williams, A. C. D. (2002). Facial expression of pain: An evolutionary account. *Behavioral and Brain Sciences*, 25, 439–88.

2 Jackson, P. L., Meltzoff, A. N. and Decety, J. (2005). How do we perceive the pain of others: A window into the neural processes involved in empathy. *NeuroImage*, 24, 771–9.

3 Masten, C. L., Morelli, S. A. and Eisenberger, N. I. (2011). An fMRI investigation of empathy for 'social pain' and subsequent prosocial behavior. *NeuroImage*, 55, 381–8; Meyer, M. L., Masten, C. L., Ma, Y., Wang, C., Shi, Z., Eisenberger, N. I. and Han, S. (2013). Empathy for the social suffering of friends and strangers recruits distinct patterns of brain activation. *Social Cognitive and Affective Neuroscience*, 8, 446–54.

4 Zadro, L., Williams, K. D. and Richardson, R. (2004). How low can you go? Ostracism by a computer is sufficient to lower self-reported levels of belonging, control, self-esteem, and meaningful existence. *Journal of Experimental Social Psychology*, 40 (4), 560–67.

5 Gonsalkorale, K. and Williams, K. D. (2007). The KKK won't let me play: Ostracism even by a despised outgroup hurts. *European Journal of Social Psychology*, 37 (6), 1176–86.

6 Sullivan, M. J. L., Martel, M. O., Tripp, D., Savard, A. and Crombez, G. (2006). The relation between catastrophizing and the communication of pain experience. *Pain*, 122, 282–8.

7 Gray, K. and Wegner, D. M. (2010). Torture and judgments of guilt. *Journal of Experimental Social Psychology*, 46, 233–5.

8 Master, S. L., Eisenberger, N. I., Taylor, S. E., Naliboff, B. D., Shirinyan, D. and Lieberman, M. D. (2009). A picture's worth: Partner photographs reduce experimentally induced pain. *Psychological Science*, 20, 1316–18.

9 Schachter, S. (1959). *The Psychology of Affiliation*. Redwood City, CA: Stanford University Press.

10 Bastian, B., Jetten, J., Hornsey, M. H. and Chong, M. (in preparation). Physical pain promotes affiliation with others.

11 von Dawans, B., Fischacher, U., Kirschbaum, C., Fehr, E. and Heinrichs, M. (2012). The social dimension of stress reactivity: Acute stress increases prosocial behavior in humans. *Psychological Science*, 23, 651–60.

12 Gailliot, M. T., Stillman, T. F., Schmeichel, B. J., Maner, J. K. and Plant, E. A. (2008). Mortality salience increases adherence to salient norms and values. *Personality and Social Psychology Bulletin*, 34 (7), 993–1003.

13 Lim, D. and DeSteno, D. (2016). Suffering and compassion: The links among adverse life experiences, empathy, compassion, and pro-social behavior. *Emotion*, 16, 175–82.

14 Gerard, H. B. and Mathewson, G. C. (1966). The effects of severity of initiation on liking for a group: A replication. *Journal of Experimental Social Psychology*, 2, 278–87.

15 Xygalatas, D., Mitkidis, P., Fischer, R., Reddish, P., Skewes, J., Geertz, A. W., . . . and Bulbulia, J. (2013). Extreme rituals promote prosociality. *Psychological Science*, 24, 1602–5.

16 Howitt, A. W. (1884). On some Australian ceremonies of initiation. *Journal of the Anthropological Institute of Great Britain and Ireland*, 432–59.

17 Whitehouse, H. (1996). Rites of terror: Emotion, metaphor and memory in Melanesian initiation cults. *Journal of the Royal Anthropological Institute*, 2 (4), 703–15.

18 Konvalinka, I., Xygalatas, D., Bulbulia, J., Schjødt, U., Jegindø, E. M., Wallot, S., . . . and Roepstorff, A. (2011). Synchronized arousal between performers and related spectators in a fire-walking ritual. *Proceedings of the National Academy of Sciences*, 108 (20), 8514–19.

19 Winslow, D. (1999). Rites of passage and group bonding in the Canadian Airborne. *Armed Forces & Society*, 25 (3), 429–57.

20 Elder Jr, G. H. and Clipp, E. C. (1988). Wartime losses and social bonding: Influences across 40 years in men's lives. *Psychiatry*, 51 (2), 177–98.

21 Whitehouse, H., McQuinn, B., Buhrmester, M. and Swann, W. B. (2014). Brothers in arms: Libyan revolutionaries bond like family. *Proceedings of the National Academy of Sciences*, 111 (50), 17783–5.

22 Penner, L., Brannick, M. T., Webb, S. and Connell, P. (2005). Effects on volunteering of the September 11, 2001, attacks: An archival analysis. *Journal of Applied Social Psychology*, 35 (7), 1333–60.

23   Bastian, B., Jetten, J. and Ferris, L. J. (2014). Pain as social glue: Shared pain increases cooperation. *Psychological Science*, 25, 2079–85.

24   Taylor, S. E., Klein, L. C., Lewis, B. P., Gruenewald, T. L., Gurung, R. A. R. and Updegraff, J. A. (2000). Biobehavioral responses to stress in females: Tend-and-befriend, not fight-or-flight. *Psychological Review*, 107, 411–29.

25   Bastian, B., Jetten, J., Thai, H. and Steffans, N. K. (in preparation). Shared pain enhances creativity by generating group cohesiveness.

6. FINDING FOCUS

1   Le Breton, D. (2000). Playing symbolically with death in extreme sports. *Body & Society*, 6 (1), 1–11.

2   Twofeathers, M. (1997). *Road to the Sundance: My Journey into Native Spirituality*. New York: Hyperion.

3   Jegindø, E. M. E., Vase, L., Jegindø, J. and Geertz, A. W. (2013). Pain and sacrifice: Experience and modulation of pain in a religious piercing ritual. *International Journal for the Psychology of Religion*, 23 (3), 171–87.

4   Ibáñez, Pedro (1882). *La Vida de la Santa Madre Teresa de Jesús*. Madrid (English translation, *The Life of S. Teresa of Jesus*. London, 1888).

5   Eccleston, C. (1994). Chronic pain and attention: A cognitive approach. *British Journal of Clinical Psychology*, 33 (4), 535–47.

6   For a review, see Eccleston, C. and Crombez, G. (1999). Pain demands attention: A cognitive–affective model of the interruptive function of pain. *Psychological Bulletin*, 125 (3), 356.

7   Fredrickson, B. L. (2001). The role of positive emotions in positive psychology: The broaden-and-build theory of positive emotions. *American Psychologist*, 56 (3), 218.

8   http://carolhortonphd.com/blog/.

9   Califia, P. (1983). A secret side of lesbian sexuality. In T. Weinberg and G. Kamel (eds.), *S and M: Studies in Sadomasochism*. Buffalo, NY: Prometheus, pp. 129–36.

10   Masters, W. H. and Johnson, V. E. (1970). *Human Sexual Inadequacy*. Boston, MA: Little, Brown.

11   Scarry, E. (1985). *The Body in Pain: The Making and Unmaking of the World*. New York: Oxford University Press.

12   Craig, A. D. (2009). How do you feel – now? The anterior insula and human awareness. *Nature Reviews Neuroscience*, 10, 59–70; see also Craig, A. D. (2002). How do you feel? Interoception: The sense of the physiological condition of the body. *Nature Reviews Neuroscience*, 3, 655–66.

13 Craig, A. D. (2003). A new view of pain as a homeostatic emotion. *Trends in Neurosciences*, 26 (6), 303–7.

14 Sorokin, P. A. (1967). *The Ways and Power of Love*. Chicago: H. Regnery (first published 1954), p. 130.

15 Terray, L. (1961). *Les Conquérants de l'inutile*. Paris: Gallimard, quoted in Le Breton, D. (2000). Playing symbolically with death and extreme sports. *Body & Society*, 6 (1), 1–11.

16 Willig, C. (2008). A phenomenological investigation of the experience of taking part in extreme sports. *Journal of Health Psychology*, 13 (5), 690–702.

17 Berkman, E. T., Burklund, L. and Lieberman, M. D. (2009). Inhibitory spillover: Intentional motor inhibition produces incidental limbic inhibition via right inferior frontal cortex. *NeuroImage*, 47 (2), 705–12.

18 Tuk, M. A., Trampe, D. and Warlop, L. (2011). Inhibitory spillover: Increased urination urgency facilitates impulse control in unrelated domains. *Psychological Science*, 22 (5), 627–33.

19 Fenn, E., Blandón-Gitlin, I., Coons, J., Pineda, C. and Echon, R. (2015). The inhibitory spillover effect: Controlling the bladder makes better liars. *Consciousness and Cognition*, 37, 112–22.

20 Bastian, B., Jetten, J. and Hornsey, M. J. (2014). Gustatory pleasure and pain: The offset of acute physical pain enhances responsiveness to taste. *Appetite*, 72, 150–55.

21 Raichle, M. E., MacLeod, A. M., Snyder, A. Z., Powers, W. J., Gusnard, D. A. and Shulman, G. L. (2001). A default mode of brain function. *Proceedings of the National Academy of Sciences*, 98 (2), 676–82.

22 Mason, M. F., Norton, M. I., Van Horn, J. D., Wegner, D. M., Grafton, S. T. and Macrae, C. N. (2007). Wandering minds: The default network and stimulus-independent thought. *Science*, 315 (5810), 393–5.

23 Killingsworth, M. A. and Gilbert, D. T. (2010). A wandering mind is an unhappy mind. *Science*, 330 (6006), 932.

24 Wilson, T. D., Reinhard, D. A., Westgate, E. C., Gilbert, D. T., Ellerbeck, N., Hahn, C., . . . and Shaked, A. (2014). Just think: The challenges of the disengaged mind. *Science*, 345 (6192), 75–7.

7. THE MEANINGFUL LIFE

1 Ryff, C. D. (1989). Happiness is everything, or is it? Explorations on the meaning of psychological well-being. *Journal of Personality and Social Psychology*, 57 (6), 1069.

2 http://ideas.time.com/2013/08/01/do-children-bring-happiness-or-misery/.

3 Baumeister, R. F., Vohs, K. D., Aaker, J. L. and Garbinsky, E. N. (2013). Some key differences between a happy life and a meaningful life. *Journal of Positive Psychology*, 8 (6), 505–16.

4 Schueller, S. M. and Seligman, M. E. (2010). Pursuit of pleasure, engagement, and meaning: Relationships to subjective and objective measures of well-being. *Journal of Positive Psychology*, 5 (4), 253–63.

5 Randles, D., Heine, S. J. and Santos, N. (2013). The common pain of surrealism and death: Acetaminophen reduces compensatory affirmation following meaning threats. *Psychological Science*, 24, 966–73.

6 Shackman, A. J., Salomons, T. V., Slagter, H. A., Fox, A. S., Winter, J. J. and Davidson, R. J. (2011). The integration of negative affect, pain and cognitive control in the cingulate cortex. *Nature Reviews Neuroscience*, 12, 154–67.

7 Gray, K. and Wegner, D. M. (2010). Blaming God for our pain: Human suffering and the divine mind. *Personality and Social Psychology Review*, 14 (1), 7–16.

8 Sibley, C. G. and Bulbulia, J. (2012). Faith after an earthquake: A longitudinal study of religion and perceived health before and after the 2011 Christchurch New Zealand earthquake. *PLoS One*, 7 (12), e49648.

9 Norenzayan, A. and Hansen, I. G. (2006). Belief in supernatural agents in the face of death. *Personality and Social Psychology Bulletin*, 32 (2), 174–87.

10 Whitson, J. A. and Galinsky, A. D. (2008). Lacking control increases illusory pattern perception. *Science*, 322 (5898), 115–17.

11 Simonov, P. V., Frolov, M. V., Evtushenko, V. F. and Sviridov, E. P. (1977). Effect of emotional stress on recognition of visual patterns. *Aviation, Space, and Environmental Medicine*, 48 (9), 856–8.

12 Oishi, S. and Diener, E. (2014). Residents of poor nations have a greater sense of meaning in life than residents of wealthy nations. *Psychological Science*, 25 (2), 422–30.

13 http://www.mrdavidzammit.com/writing/2016/5/22/6covoovzn1nwyw 1doow54b9jkcckqu.

14 Gillman, P. (ed.) (2010). *Climbing Everest: The Complete Writings of George Mallory*. London: Gibson Square Books.

15 Olivola, C. Y. and Shafir, E. (2013). The martyrdom effect: When pain and effort increase prosocial contributions. *Journal of Behavioral Decision Making*, 26 (1), 91–105.

16 King, L. A., Hicks, J. A. and Abdelkhalik, J. (2009). Death, life, scarcity, and value: An alternative perspective on the meaning of death. *Psychological Science*, 20 (12), 1459–62.

17 Vess, M., Routledge, C., Landau, M. J. and Arndt, J. (2009). The dynamics of death and meaning: the effects of death-relevant cognitions and personal need for structure on perceptions of meaning in life. *Journal of Personality and Social Psychology*, 97 (4), 728–44.

18 Ryan, R. M. and Deci, E. L. (2000). Self-determination theory and the facilitation of intrinsic motivation, social development, and well-being. *American Psychologist*, 55 (1), 68.

CONCLUSION

1 https://www.psychologytoday.com/blog/compassion-matters/201203/is-our-aversion-pain-killing-us.

2 http://www.cdc.gov/injury/pdfs/budget/pdo.pdf.

3 http://apps.who.int/medicinedocs/documents/s19032en/s19032en.pdf.

4 *The Notebooks of Leonardo da Vinci*, trans. Edward MacCurdy (1939). Garden City, NY: Garden City Publishing, p. 1097.